Supporting Rural

Praise for this book...

'Impressive gains made in the past few decades to increase water supply coverage in support of global and national MDG targets are at risk. Low sustainability of rural water supply systems and poorly functioning community-based management groups are a challenge for communities, national governments and donors. *Supporting Rural Water Supply* provides a much-needed evidence base on what works and a practical approach to planning a more sustainable future based on a service delivery approach.'
         Marcus Howard, *Water Adviser, Australian Agency for International Development*

'This book represents a critical milestone in our understanding of the rural water sector – drawing together as it does a wide range of experiences across many different country contexts. The analysis not only unpacks some of the most important and persistent problems facing the sector, but also provides clear guidance for both policy makers and practitioners on the possible ways forward.'
         Richard Carter, *Head of Technical Support, WaterAid, UK*

'A must-read for those supporting progress in the water sector.'
         Monica Ellis, *CEO, Global Water Challenge*

'This study looks at how to ensure that rural water supplies, once constructed, continue to provide safe water to their consumers. Sustainability is a key issue facing the sector. A summary of lessons learned in this respect deserves our full attention.'
         Elizabeth Kleemeier, *Senior Water and Sanitation Specialist, World Bank*

'A must for all practitioners and decision makers striving to ensure sustainable drinking water security in rural communities around the world.'
         Christophe Prevost, *Senior Water and Sanitation Specialist, Water and Sanitation Program – South Asia, World Bank*

Supporting Rural Water Supply
Moving towards a Service Delivery Approach

Harold Lockwood and Stef Smits

Practical Action Publishing Ltd.
The Schumacher Centre
Bourton on Dunsmore, Rugby,
Warwickshire, CV23 9QZ, UK
www.practicalactionpublishing.org

© IRC International Water and Sanitation Centre and Aguaconsult, 2011

ISBN 978 1 85339 729 5

This publication is copyright, but may be reproduced by any method without fee for not-for-profit educational, scientific or development-related purposes, but not for resale. Formal permission is required for all such uses, but normally will be granted immediately. For copying in any other circumstances, or for re-use in other publications, or for translation or adaptation, prior written permission must be obtained from the publisher.

A catalogue record for this book is available from the British Library.

The authors have asserted their rights under the Copyright Designs and Patents Act 1988 to be identified as authors of this work.

Since 1974, Practical Action Publishing (formerly Intermediate Technology Publications and ITDG Publishing) has published and disseminated books and information in support of international development work throughout the world. Practical Action Publishing is a trading name of Practical Action Publishing Ltd (Company Reg. No. 1159018), the wholly owned publishing company of Practical Action. Practical Action Publishing trades only in support of its parent charity objectives and any profits are covenanted back to Practical Action (Charity Reg. No. 247257, Group VAT Registration No. 880 9924 76).

Cover photo credits:
Left: Men discussing water supply. Image courtesy of WEDC © Sue Coates
Right: Women carrying water, Luena, Angola. Image courtesy of WEDC
© Wayne Conradie (Picturing Africa)
Cover design by Practical Action Publishing
Indexed by Andrea Palmer
Typeset by Practical Action Publishing
Printed by Hobbs the Printers Ltd
Printed on FSC 100% post-consumer waste recycled paper

Contents

Figures	viii
Tables	ix
Boxes	x
Acknowledgments	xi
About the authors	xii
Executive summary	**1**
1. Introduction	**11**
Background	11
The objectives of Triple-S	12
Structure of this book	13
2. Methodology and conceptual framework	**15**
Country selection and context	15
Methodology and common analytical framework	16
The analytical tool	17
Key concepts and terminology	18
The Service Delivery Approach and Service Delivery Models	18
Institutional levels and functions	20
Water supply service levels	22
Sustainability	23
Scale and scaling up	24
Cost categories for the full life-cycle of rural water services	25
Decentralisation	25
Aid effectiveness	27
3. Country sketches	**29**
Benin	30

	Burkina Faso	32
	Colombia	34
	Ethiopia	36
	Ghana	38
	Honduras	40
	India	42
	Mozambique	44
	South Africa	46
	Sri Lanka	48
	Thailand	50
	Uganda	52
	United States of America	54
4.	**Findings from the country studies**	57
	The status of access to sustainable rural water supply services	58
	Service levels	58
	Sustainability definitions, indicators and targets	60
	Discussion: indicators for sustainable service delivery	65
	Institutional arrangements for rural water supply: sector reforms separating roles and functions and decentralisation	65
	Decentralisation processes	66
	Reform to functions in rural water service delivery	68
	Reforming agencies	70
	Discussion: impact of decentralisation and reforms on conditions for service delivery	72
	Management options as part of Service Delivery Models	73
	Community-based management	75
	Beyond CBM: private sector management arrangements and rural utilities	81
	Self-supply	83
	Discussion: defining Service Delivery Models as part of the Service Delivery Approach	85
	Service authority functions	88
	Planning and implementation of water services	88

	Accountability, regulation, oversight and enforcement	94
	Monitoring and information management	99
	Post-construction support	103
	Discussion: authority functions to facilitate service delivery	110
	An enabling environment for service authorities	111
	Policy and strategy development	112
	Financing sustainable service delivery	113
	Learning and sector capacity building	122
	Aid effectiveness: harmonisation and alignment	128
	Discussion: supporting the Service Delivery Approach	134
5.	**Conclusions and recommendations**	**137**
	A spectrum of approaches to rural water supply	138
	Factors in moving towards the Service Delivery Approach	142
	Recommendations for the sustainable provision of rural water services at scale	147

References	**153**
Annexes	**161**
Annex A: List of abbreviations, acronyms and non-English terms	161
Annex B: Glossary	167
Annex C: Analytical tool for country studies	170
Annex D: Triple-S principles framework	176
Annex E: List of country reports and literature studies	178

Figures

Figure 1:	Stages in the service delivery cycle	18
Figure 2:	Water service delivery from the user perspective: repeated disappointment, or a Service Delivery Approach	20
Figure 3:	Service Delivery Models	22
Figure 4:	Service levels	23
Figure 5:	Main component and terminology related to aid effectiveness	28
Figure 6:	Estimated % of broken handpumps	64
Figure 7:	Arrangements for letting and regulating delegated contracts by the WSAs in South Africa	84
Figure 8:	Service Delivery Models and types of settlement	87
Figure 9:	Service delivery life-cycle in South Africa	92
Figure 10:	The framework of accountability relationships	95
Figure 11:	Process for support to WSAs in South Africa	124
Figure 12:	The continuum from implementation towards Service Delivery Approaches	141
Figure 13:	The evolution of the rural water sector	142

Tables

Table 1:	Study countries and some basic development sector indicators	15
Table 2:	Definitions of cost categories	26
Table 3:	Dimensions and modes of decentralisation	27
Table 4:	Access to rural water supply services	59
Table 5:	(Proxy) indicators in use for sustainability of rural water supply	60
Table 6:	Formally recognised Service Delivery Models	74
Table 7:	Ghana: formally recognised sub-models of CBM	77
Table 8:	Service Delivery Models and variants	86
Table 9:	Uganda's eleven golden indicators	101
Table 10:	Generic characteristics of post-construction and capacity support	104
Table 11:	Post-construction and capacity support	105
Table 12:	Comparison of costs of post-construction support in Colombia	118
Table 13:	Donor dependency in the water and sanitation sector in selected study countries	121
Table 14:	Recommendations for shifting from implementation to Service Delivery Approaches	150

Boxes

Box 1:	Categorisation of sustainability of rural water supply systems in Honduras	62
Box 2:	Change management works!	71
Box 3:	Colombia's entrepreneurial culture programme	79
Box 4:	Different approaches to delegation in Benin	82
Box 5:	Promoting rainwater harvesting to support self-supply approach in Thailand	84
Box 6:	Shared management based on 'mutualisation' of costs in a broader service area in Burkina Faso	115
Box 7:	Tracking sector investments in the joint performance report in Uganda	120
Box 8:	Uganda's Technical Support Units (TSUs)	123
Box 9:	The Honduran network for water and sanitation, RASHON	126
Box 10:	Interview results from Ethiopia on the national One WASH Programme	129
Box 11:	Tamil Nadu – a multiplicity of agencies at the village level	133

Acknowledgements

We gratefully acknowledge the writers and contributors to the individual country reports and literature reviews which form the building blocks of this global synthesis; without their in-depth investigations, questioning and analysis it would not have been possible to write this book. They are:

Study:	Authors:
Benin	Cyriaque Adjinacou, MGE Conseils, Benin
Burkina Faso	Denis Zoungrana, Institut International d'Ingénierie de l'Eau et de l'Environnement (2IE), Burkina Faso
Colombia	Johnny Rojas Padilla, Adriana Zamora Trejos, Paola Tamayo Andrade and Mariela Garcia Vargas, CINARA, Colombia
Ethiopia	Tamene Chaka, Leulseged Yirgu and Zemede Abebe, RiPPLE, Ethiopia; and John Butterworth, IRC International Water and Sanitation Centre, The Netherlands
Ghana	Tania Verdemato, Aguaconsult, United Kingdom; Jean de la Harpe, Marieke Adank and Patrick Moriarty, IRC International Water and Sanitation Centre, The Netherlands; Mawuena Doste, Maple Consult, Ghana; Benedict Tuffuor, TREND, Ghana; and Nana Boachie Danquah, University of Ghana, Ghana
Honduras	Manuel A. López, independent consultant, Honduras
India	A. J. James, ICRA Management Consulting Services Limited, India
Mozambique	Carlos Gideon José Munguambe and Vitoria Afonso Langa de Jesus, Cowater, Mozambique
South Africa	Jean de la Harpe, IRC International Water and Sanitation Centre, The Netherlands
Sri Lanka	A. J. James, ICRA Management Consulting Services Limited, India
Thailand	Muanpong Juntopas and S. Naruchaikusol, Stockholm Environment Institute, Asia Centre, Thailand
Uganda	Cate Nimanya, Harriet Nabunnya and Solomon Kyeyune, NETWAS Uganda; and Han Heijnen, independent consultant, Uganda
USA	Stephen Gasteyer, Department of Sociology, Michigan State University, United States of America

Harmonisation and alignment literature review	Jean de la Harpe, IRC International Water and Sanitation Centre, The Netherlands
Service delivery concepts literature review	John Butterworth, IRC International Water and Sanitation Centre, The Netherlands

We would like to recognise the inputs of Christophe Prevost of the Water and Sanitation Program South Asia at various points in the development of this book, whose insights and support have been much appreciated. We are indebted to Dr. Patrick Moriarty, Regional Coordinator for West Africa, IRC International Water and Sanitation Centre, and Dr. Richard Carter, Head of Technical Support, WaterAid, United Kingdom both of whom made invaluable contributions and comments on earlier drafts of this book. We thank Kerry Harris, Development Consultant, South Africa for her work on editing the book.

Harold Lockwood and Stef Smits
The Hague, 2011

About the authors

Harold Lockwood is an expert in water supply and sanitation with over twenty years of international experience. His particular areas of interest are institutional analysis, sector reform and policy development; decentralisation of service provision; and sustainability issues relating to the management of rural water services. He has worked extensively in Latin America, Asia and Africa and has served as technical advisor to Pakistan's Local Government and Rural Development Department and to Nicaragua's National Institute for Water and Sanitation. Harold is the director of the UK consulting firm Aguaconsult and regularly undertakes field assignments for a broad range of clients including major bi-lateral donors, multi-lateral agencies, UN organizations, international NGOs, Foundations and private sector companies. He is currently working on Triple-S, a global initiative to improve sustainability of rural water services.

Stef Smits is a programme officer at IRC International Water and Sanitation Centre in The Netherlands. His main area of interest is rural water supply, and particularly the sustainability of services. He is currently the secretary of the MUS Group, a community of practice dedicated to work on multiple-use services. Stef coordinates IRC's Latin America Regional Programme and leads the Triple-S research work. He has experience in Latin America – with a focus on Honduras, Colombia and Bolivia – and in Southern Africa, where he has worked primarily in South Africa and Zimbabwe.

Executive summary

Collectively, billions of dollars have been invested in the provision of rural water supply systems in developing countries over the past three decades. This period has also seen an evolution in thinking and practice around the approaches to delivering water supply to rural populations. We have moved from supply-driven centralised government programming to more demand-driven approaches, based on the philosophy of community participation with community-based management emerging as the principal management vehicle from the 1980s onwards in most countries. In more recent years there has been a call to build on community management with more structured systems of post-construction support and the increasing involvement of local private operators. Global monitoring results tell us that progress is being made and that even including population growth, we are increasing the rate of coverage in many, but not all, countries at a pace that will meet the Millennium Development Goals.

However, the reality behind these aggregated figures is often quite different in terms of the access to a service that users actually experience: communities unable to cope with management of their schemes, poor maintenance, lack of financing, breakdowns, poor water quality, lack of support and, ultimately, an unreliable and disrupted supply of water to households. Commonly cited figures from a range of countries put non-functionality at somewhere between 30-40% of all systems at any one time. While it can be argued that the reverse is also true – that in fact two-thirds of community-managed systems still function somehow – tolerance of this level of failure would not be contemplated in most other spheres of public service. Clearly something has not been working as it should. Where there have been successes – and there have been many specific cases – these tend to require highly intensive inputs (often from donor-driven programmes or non-governmental organisations) or are simply the result of strong, motivated individuals within communities that have the energy and leadership to make things work. Many of these successes have been documented, but they remain largely as islands, without the possibility of being replicated systematically and eventually scaled up.

This study takes a critical look at why we have been unable to provide a sustainable water service to rural people for so long. It seeks to identify some of the most important factors that appear to contribute to, or constrain, the delivery of such services. We do this by drawing upon a series of case studies from 13 countries which were carried out as part of a global learning initiative to contribute to improved water services – Sustainable Services at Scale, or Triple-S. The countries were selected to represent a range of socio-economic contexts and aid dependency, as well as relative development of the water sector, from those with ongoing reform and decentralisation efforts to more 'mature' sectors where such processes have been established for some time.

Building blocks towards more sustainable service delivery	
Professionalisation of community management	Community management must be properly embedded in, and supported by, policy, legal and regulatory frameworks and support services, both at national and local levels; in order to become more effective community-based management entities must be legally recognised.
Increased recognition and promotion of alternative service provider options	There is a range of different management options beyond community-based management – including local private operators – that can better support different service levels, technology and types of settlements; these should be described and set out in clear Service Delivery Models which are well disseminated.
Sustainability indicators and targets	Monitoring and target indicators should move beyond systems built and 'beneficiaries' served and include benchmarking against the services delivered and the performance of service providers.
Post-construction support to service providers	Most community-based management and local private operators cannot manage on their own; there is a need for structured systems of support that are properly funded to back-up and monitor these service providers; in many cases it is local government that will take on this responsibility.
Capacity support to decentralised government (to the service authorities)	Many local governments will require help and support if they are going to fulfil their role in guaranteeing services; ongoing capacity support programmes covering key functions in the life-cycle of rural water supply services, including management, procurement and contracting are needed and must be paid for.
Learning and sharing of experience	Learning and knowledge management is an important element of any mature sector; this should not rely on *ad hoc* support, but become an integral part of sector capacity and be properly funded both at national and decentralised levels.
Planning for asset management	One of the main weaknesses of rural water provision has been the lack of proper asset management; systematic planning, inventory updates and financial forecasting should be introduced; ownership of assets must be better defined so as to allow for delegated contracting, where appropriate.
Financial planning frameworks to cover all life-cycle costs	Sector financial frameworks must be expanded beyond the basics of capital investments and minor operation and maintenance costs; all life-cycle costs must be accounted for, especially major capital maintenance expenditure, and direct and indirect support costs.
Regulation of rural services and service providers	Service provision – and the performance of service providers of all types – should eventually be regulated, even where this is done with a 'light touch' system. Any attempts to establish regulation should apply appropriate performance criteria and not be overly punitive for fledgling rural operators.

Summary findings

The study has revealed a spectrum of approaches to rural water supply. We can classify these on a continuum from what can be identified as a largely infrastructure or 'implementation' focus at one extreme, to Service Delivery Approaches at the other. In simple terms, the former typifies sectors in which structures, systems and efforts are mainly geared towards the capital-intensive phase of rural water (i.e. constructing facilities and initial implementation and training), whereas the latter is geared towards a more balanced attention to the full life-cycle of a service, including aspects such as post-construction support, investment planning for longer-term capital maintenance and asset renewal, and a learning and adaptive sector.

In many of the case study countries it is evident that the rural water supply sector has gone, or is still going through, a process of change as broader decentralisation efforts continue to be implemented; and the water sector itself has been subject to reforms. Chief amongst these reforms is the separation of functions previously held by mainly centralised state agencies including planning, operation and regulation (even though in the rural water sector, regulation is still either non-existent or nascent at best) of services. On the back of decentralisation of public services, a new and much more substantive role for local government has been established, which includes important functions for guaranteeing water for rural populations at local level; in this analysis we refer to these as 'service authority' functions. However, in a number of cases the hoped for reforms – and clear separation of functions – exist largely on paper, with the reality being that not much has changed in practice. This can be due to a range of reasons, including of lack of capacity at the local level, or a certain degree of inertia and even resistance on the part of strong parastatal central government organisations to devolve authority and resources.

As part of these processes of reform, there are a number of factors that appear to be common in adopting sector changes toward a more service-oriented approach. Some are more incipient than others; some are still only changes in discourse at this stage and not yet changes in practice, and there are obvious differences between countries. These changes are seen to be happening at three key levels.

Service provider level

The findings of the various country studies point to a taxonomy of models at the level of the service provider, which comprises four main options, namely: community-based management, direct public sector provision, private sector operators and self-supply. In this study we consider service providers to include community-management entities and local private operators, as well as publicly mandated utilities, and individual households. In reality there are a number of variants within most of these categories, as well as hybrids, reflecting different degrees of system complexity and levels of service demanded.

While there is no compelling evidence of any 'right' or 'wrong' Service Delivery Models, what is clear is that there has been a generalised trend away from the more voluntary arrangements of community-based management, towards professionalisation, or what has been termed *'community management plus'*. Some of the founding principles of community-based management, such as community cohesion, common participation in the broader user community and informal accountability to a water committee, are crucial though insufficient. The ideals behind it are often undermined by lack of formalisation of these arrangements within broader local government bye-laws and national legislation and policy, the absence of clear contracting, lack of legal standing of the committees, and the lack of professional capacity in certain aspects of running and managing systems. In the few cases from the study countries where community-management approaches have worked well at scale (e.g. the Water and Sanitation Management Organisation experience from Gujarat State in India), these have been based on explicit efforts to formalise community-based management within local government structures. Even with the drive to professionalise community-based management, it is likely that more conventional voluntary arrangements will still predominate in smaller systems, often with a point source supply providing lower levels of service to consumers.

A second and related trend illustrated by the studies is that there is an emerging but growing role for small-scale private operators to improve service provision for rural populations. Population growth and higher density rural growth settlements are making the distinction between the demands and solutions of these populations and the truly low-density rural villages and hamlets increasingly clear. Demand for higher levels of service, better and more competent management, and increased opportunities for revenue collection all point towards a more professionalised service. Although these approaches only account for a relatively small proportion at present, the involvement of private operators in public-private partnerships is growing and appears to be a good fit for larger, more complex piped systems, providing higher service levels in larger rural communities or growth centres.

Finally, self-supply is recognised as an important approach in some countries, while it is ignored in others. In many places, particularly in highly dispersed rural settlements, self-supply and household management is happening in a *de facto* manner. However, more gains can be made if this is recognised as a Service Delivery Model and supported as such, as has been done, for example, in Thailand.

Service authority level

Adopting a Service Delivery Approach means having systems and capacity in place at the decentralised level to support different service providers in effectively operating, administering and maintaining rural water systems. Under decentralisation, far greater responsibility has been transferred to local or district government; we refer to this as the 'intermediate' level of government.

Results from the country studies present a range of scenarios that illustrate how these functions are understood and provided for in practice. There are situations where many of these functions are still geared towards the capital intensive part of the life-cycle, through delivery of new or rehabilitated hardware and corresponding software activities (such as establishing and training water committees). In this scenario many service authority functions focus on immediate outputs: that is to say, planning processes are primarily concerned with new physical construction, with little consideration for asset management; monitoring is focused on progress in construction; and post-construction support, if considered at all, consists of a few months follow-up after completion of the project. Not surprisingly, this mode of water supply provision is most prominent in those countries where coverage is low, and the whole sector is geared towards increasing coverage mainly through implementation of new systems, for example, in Ethiopia or Mozambique.

The other end of the scale shows examples of these service authority functions with a much greater emphasis on service *delivery*. This is typified by planning that covers the implementation of projects, long-term asset management and support to existing systems; the monitoring of progress in outputs and of the service provided (including aspects of the performance of the service provider); and that dedicates a greater proportion of resources to post-construction support. This scenario tends to be the case in countries where coverage figures are reasonably high and where the performance and quality of service has become a concern, rather than simply driving up coverage rates. Examples include Colombia, USA, South Africa and Thailand.

Of course, these two scenarios represent extremes of the spectrum, and many countries show a mixed picture with better progress in some areas than in others. There are also examples of 're-centralisation' of certain functions through the creation of groupings of local government, or municipalities (or *'mancomunidades'* as they are known in Latin America) which share the costs of specialised technical inputs, and can reach greater economies of scale than individual local government authorities. In almost all cases, the fulfilment of service authority functions is hampered by lack of adequate financing and technically qualified staff at the intermediate level. This capacity problem is a critical one and goes beyond the confines of the water and sanitation sector, being part of a much broader set of issues around public administration delivery and fiscal decentralisation. Nonetheless, providing support to local government capacity building is an area that is starting to be addressed more systematically in countries such as South Africa, Uganda and some states in India, but remains very weak in others, such as Mozambique and Burkina Faso.

National-level enabling environment

The case studies illustrate both positive and negative examples of how sector policies and legislative frameworks have been clarified to support improved service provision. One of the most commonly cited success factors is the presence of a strong national vision or strategy and catalytic (political and bureaucratic) champions. Experiences from India, Uganda, South Africa and Thailand all show how having such clear leadership can accelerate the process of change and support the definition of clear policies and institutional frameworks. But even in cases where the enabling environment is relatively well established, there are often challenges around dissemination of roles and responsibilities, and a lack of harmonisation and coordination.

The findings from the studies highlight the fact that for scalability of approaches to work well, there needs to be strong policy, planning, investment decisions and capacity building done at a sector level, with the goal of setting clear frameworks for all actors to function within. This type of scalability is associated with those countries that have a relatively high degree of harmonisation and alignment of sector organisations to national policies and frameworks. The converse is a situation where the water sector is fragmented, investments are made in a 'projectised' way, and structural support for improved capacity is either weak or absent. Such fragmentation was seen to be problematic in countries such as Burkina Faso, Ethiopia and Honduras. The results are poorly defined policy, gaps in legislation, the absence of structured support to decentralised level actors, and confusion about who is responsible for what, particularly regarding financing different types of expenditure. Not surprisingly, the resultant vacuum tends to be filled by a plethora of 'implementers', many of whom follow their own approaches and policies, simply reinforcing the confusion and lack of commonly agreed 'rules of the game'.

One factor, or driver, that emerges consistently across almost all of the country case studies is what can be termed the 'political economy' of rural water. By this we refer to the politics of donor power, of government decision-making and political interference – both from the centre and in the local politics of decentralised government – and the associated drive to control resource allocation, as well as corruption and nepotism. Across the many different contexts the formal development of the rural water sector takes place against a complex backdrop of powerful interests, competing agendas and dynamics, many of which are never formally captured in sector documentation or evaluations. Taken on aggregate, these forces can often reinforce an emphasis on capital investment (in water systems) for financial or political gain, and conversely undermine an emphasis on Service Delivery Approaches which are less expedient. But political engagement can also be engineered for the good with (some) examples of champions, both politicians and senior and influential civil servants, who can support change processes and help to drive through complex reforms.

Capital investment costs and minor operation costs are relatively well defined within most national policy and macro-level financial planning processes. However, very few countries specify the financing requirements for three other critical components, namely: large-scale capital maintenance and replacement expenditure, and indirect and direct support costs, including the vital function of post-construction support and monitoring. In cases where this is done in a more systematic way, such as South Africa and Uganda, the resources made available are insufficient to deal with the maintenance backlog. Asset management planning, which is a relatively common tool for urban utilities, is practically unknown in the rural water sector.

Following the broader trend towards improving aid effectiveness, bi-lateral and multi-lateral development partners are increasingly working within government-led national sector frameworks. But the picture is still mixed, and there are cases where donors (including international non-governmental organisations and charities) continue to work outside of national guidelines. For example, Uganda, despite having one of the longest standing sector-wide approaches in the water sector, is still subject to a significant proportion of investment by non-state actors which by-pass common funding mechanisms and which do not comply with the norms and standard approaches set by governments. The negative impact of such continued fragmentation depends on the level of aid-dependency, the strength of government vision and leadership, and the relative mix of external donor and non-governmental organisation activity. What is clear from the studies is that greater levels of harmonisation and alignment – often involving sector-wide approaches or sector basket funding mechanisms – can better facilitate structured investments in strengthening sector systems. This, in turn, allows for a stronger enabling environment at national level, and improved capacity at local government level.

Implications for policy and development partner assistance

The emerging picture with regard to rural water sector development, although mixed overall, is one of a general shift, or attempt to shift, towards a Service Delivery Approach and away from a focus on infrastructure. Many elements of this shift are manifested in the various case studies but, most significantly, this trend marks the establishment of *sector wide* frameworks that transcend any one (donor) programme and establish clear roles and legal frameworks. In this way a multiplicity of actors can be enabled to work within known 'rules of the game' so as to support sustainable provision of a service of a given type and level. This is the logical conclusion of working at scale. This does not necessarily imply a 'one-size fits all' national programme; rather, by providing clear policies, guidelines and norms, an appropriate mix of management options can be adopted by different agencies to meet the demands of a range of populations and service levels.

We all share a vision to see better and more reliable water services supplied to rural populations, and to see the benefits of investments sustained over as long a time period as possible. But what recommendations can we give that can promote the adoption of service-oriented approaches that can operate at scale? Based on the findings of the studies, we can identify a number of building blocks which can contribute to supporting this shift towards the delivery of services; these building blocks are summarised in the table on page 2.

Promoting Service Delivery Approaches across different contexts

These recommendations would have to be applied on a country-specific basis, and we can already differentiate between countries which are still struggling to meet the basic challenge of increasing coverage, and those that are moving towards so-called 'second generation' problems once a critical mass of coverage has been achieved. In considering the policy implications for how to tackle this problem, we have identified three broad sets of scenarios with regard to sector development:

- Firstly, those countries in which coverage is really still very low (for example, Ethiopia or Mozambique from our study countries). It is an understandable strategy to largely focus on increasing coverage, but this should be as far as possible in a scaled-up manner. While much energy and resources go into building new systems, there may have to be an acceptance of a case of 'two steps forward, one step back' where levels of functionality will remain problematic. However, that is not to say we should not propose measures that can be taken to build the foundations for a future shift towards more service-orientated approaches.
- Where community-based management is the mainstay Service Delivery Model it should be strengthened and incorporate changes flagged by this book, particularly in the legal recognition of committees and formalising their relationships with local government. Post-construction support must be adequately addressed and financed, something that has been consistently under-funded to date. Development partner assistance to countries in this group should focus on improving alignment of programmatic support, particularly around implementation approaches, to avoid fragmentation and conflicting policies for communities.
- Secondly, we have the group of countries where coverage is already relatively high, reaching levels of 80% or more (e.g. India, Thailand, USA and Sri Lanka among our study countries), which should focus very strongly on investing in systems and capacities that underpin a true Service Delivery Approach. Such steps would include developing asset management planning, providing structured capacity support to local government, ongoing post-construction support, and financial mechanisms such as rotating funds to meet capital maintenance costs, improving life-cycle cost analysis and regulation. Another important step would be to develop specific strategies to reach the last 10-15% of the unserved populations, for example, by formally recognising self-supply, and introducing measures to support this approach in a systematic

way. Developing asset management strategies and tools for rural water services would mark another major step-change in these countries.
- Lastly, we have the middle band of countries where coverage is somewhere between 50-70% and expanding, but where there is also the very strong risk of slippage of functionality rates. This group of countries includes examples such as Honduras, Colombia, Ghana, Uganda, South Africa and Burkina Faso. These countries face an in-built tension between pursuing increased coverage (with inadequate budgets and growing populations) while simultaneously addressing sustainability in a more structured way. More capital investment is needed for new systems, extensions and higher levels of service in existing systems; but, equally, increasing attention needs to be paid to asset management, to improving management options, and monitoring and oversight of services delivered. So how should this group of countries juggle all of these balls at once?

This is indeed a critical question and the simple answer is that these middle groups must, in fact, juggle competing priorities as part of this phase of sector development. We cannot give an estimate of how much should go into new investments and how much into asset management or, in other words, what is required to build (new) systems and what is required to build and maintain the sector capacity to support the delivery of services, but what is clear is that for many countries the balance has to shift in favour of the latter. This should also be a strong message for development partners when considering funding strategies. Having a strong scaled-up implementation approach (such as the sectors adopted in South Africa, Uganda and India when they were themselves climbing the lower levels of coverage) has helped in making the transition to a Service Delivery Approach in subsequent phases.

Therefore, both national governments and development partners should invest more in building the systems of the sector to cope with the transition to service delivery, including support to professionalising community-based management, capacity support to decentralised government sector staff, and clarifying the legal and institutional frameworks for asset management and delegated contracting. Setting up and streamlining financial mechanisms, and the introduction of pooled funding, would allow for support to these types of efforts to improve the carrying capacity of the sector. Overall, development partners should take the long view with this group of countries, and move from the two to three year support horizons to much longer-term, more stable funding support which will allow for sector development in a more predictable way.

Implications for sector change

In whichever of the three groups of countries one falls, the cases also point towards three important underlying lessons for promoting more service-oriented outcomes. These are a basis for any sector change process:

Firstly, there is a tension between the need for broad-based systemic change and the practicalities of gradual improvements in specific areas. Attempting to make changes through isolated projects and programmes, by setting up stand-alone solutions, will have limited impact. Yet, achieving systemic change is not straightforward and, in many cases, one has to start pragmatically, addressing one or more of the building blocks. Frameworks, such as those used in this study, have the potential to identify specific areas or priorities, while keeping the inter-connectivity of different elements, thereby avoiding the trap of reducing sustainability issues to one or two key factors.

Secondly, in order to achieve systemic change there must be a base level of harmonisation and coordination between different actors working in the sector; this is particularly the case for the more aid-dependent countries, but it is also a factor in intra-government relationships. Working through more harmonised approaches will be key to addressing and financing systemic changes.

Finally, one has to recognise that changes in approaches to rural water reflect profound political choices, and that one has to embrace the politics of it all. We would argue in favour of change processes which are strongly vested in the political agendas, both nationally and locally, of all actors involved. Change processes, therefore, need to be accompanied by and embedded in political engagement activities.

CHAPTER 1
Introduction

Background

During the past two to three decades there has been relative success in providing new rural water infrastructure – building the physical systems – and driving increased coverage levels. However, despite this positive trend, there has to a large extent been a failure to achieve sustainable solutions. Tens of millions of rural people face continuing problems with systems that fail prematurely, leading to wasted resources and false expectations. For many of those who supposedly already enjoy an improved service, the reality is one of poor continuity, poor quality and premature failure.

Between 1990 and 2006, the absolute number of un-served people across 19 sub-Saharan African countries increased from 29 million to 272 million (RWSN, 2009). In part this is due to population growth, but many of those who supposedly count as having been 'served' actually have systems that are now not working properly or have failed completely. Both population expansion and migration patterns have led to more urbanisation, but also an increase in more densely populated rural villages or rural growth centres, with accompanying increased demand for higher levels of service. However, it is still the rural population that continues to suffer most from poor services; the Joint Monitoring Program (JMP) reports that 84% of people without access to improved drinking water sources live in rural areas (WHO/UNICEF, 2010).

Already in the early 1990s, estimates suggested that at any given moment, 30–40% of rural water supply systems in developing countries were not working (Evans, 1992). This rate has not changed much since then and although figures vary, studies from different countries indicate that somewhere between 30% and 40% of systems, particularly handpumps, still either do not function at all or are working at sub-optimal levels. The Rural Water Supply Network indicates an average rate of 36% non-functionality for hand-pumps in sub-Saharan Africa (RWSN, 2009). A more recent study by WaterAid in Tanzania indicates that only two years following installation 25% of systems are already non-functional (Taylor, 2009). Failures on this scale represent significant levels of wasted investment, probably many hundreds of millions of dollars over the last 20 years. In gravity-fed piped systems the issue is often not full collapse of services, as they are technically less prone to become fully non-functional, but providing services well below the expected performance level.

Poor sustainability of rural water supplies has been recognised for some time, and a number of management approaches have come and gone with the aim of addressing these problems; the predominant model of community management has been adopted as formal sector policy in many countries. At the same time, most efforts and resources in the Water, Sanitation and Hygiene (WASH) sector continue to go into the construction of new infrastructure, which undoubtedly is needed. However, such investment often appears to be at the expense of the sustainability of services already in place. A tipping point may now have been reached with more and more national governments and development partners beginning to recognise the scale of the problems associated with poor sustainability and the real threat this presents to achieving the WASH Millennium Development Goals (MDGs).

The objectives of Triple-S

It is against this backdrop that the Sustainable Services at Scale (or Triple-S) initiative was developed. Started in late 2008, this six-year learning initiative has the overall goal of contributing to improved sustainability of rural water services, and bringing about greater harmonisation through increased sector capacity. Triple-S aims to act as a catalyst for transforming the approaches in rural water supply from one focused on the implementation of water systems, to the provision of indefinite and reliable rural water services delivered at scale. The initiative is managed by IRC International Water and Sanitation Centre (IRC) in the Netherlands, and works in partnership with international, national and local partners[1], working initially in two focus countries – Ghana and Uganda – and expanding to Burkina Faso in 2011.

As part of the initiative's start-up, a research study was conducted between late 2009 and the second quarter of 2010. The main objective of the research study was to review and better understand the trends in rural water supply, and to identify factors that appear to contribute to or constrain the delivery of sustainable rural water services at scale. The study also sought to identify organisational incentives and barriers that shape the way in which sector institutions approach rural water services. The study was carried out in 13 countries across the globe (see Annex E for the full references of all country studies). In addition, literature reviews on into rural service provision more broadly (Butterworth, 2010) and on aid harmonisation at national and sub-national levels (de la Harpe, 2011a) were carried out. Finally, this book has been informed by a symposium on sustainable services at scale held from 13-15 April 2010 in Kampala, where many of the experiences reported here were presented and discussed, alongside work by other sector practitioners and researchers in the field (see Smits et al., 2010 for the proceedings; and Moriarty and Verdemato, 2010 for the discussion report).

As well as carrying out the studies and collation of views and information, there was an explicit goal of providing follow-up to the research in a number of countries through projects and activities of IRC and partners, with a view to generate further sector debate and action in the area of improving sustainability of services.

Structure of this book

This book presents a synthesis of the 13 country study findings and the literature reviews, and is structured around five principal sections as follows:
- Chapter 1 provides an introduction.
- Chapter 2 sets out the methodologies adopted in the country studies, and explains underlying concepts and terminologies used in the analysis of the findings.
- Chapter 3 provides a brief summary of the state of the rural water sector in each of the 13 countries.
- Chapter 4 presents the main findings from a comparative analysis across the country studies, addressing the current status of sustainability, progress on decentralisation and sector reform, the definition of Service Delivery Models (SDMs) and experiences with professionalisation of community management. This chapter also looks at the difficult question of understanding costs and financing flows. Planning, accountability, learning and capacity building are all investigated in this chapter, as well as processes of organisational change as part of sector reforms.
- Chapter 5 contains a discussion on the implications of the study findings, and conclusions are drawn up regarding the state of sustainability of rural water services. This chapter also indentifies key policy implications and provides some concrete recommendations for moving the sustainability debate forward, and for working at scale.

Endnote

1. International partners include SNV Netherlands Development Organization, Aguaconsult in the United Kingdom, the Community Water and Sanitation Agency in Ghana, and the Directorate of Water Development, Ministry of Water and Environment in Uganda. Further details can be found at: www.waterservicesthatlast.org.

CHAPTER 2
Methodology and conceptual framework

Country selection and context

Studies into the rural water sector were carried out in 13 countries with a deliberately broad range of country profiles and levels of sector development. The study countries are: Benin, Burkina Faso, Colombia, Ethiopia, Ghana, Honduras, India (three states of Gujarat, Maharashtra and Tamil Nadu), Mozambique, South Africa, Sri Lanka, Thailand, Uganda and the United States of America (USA)[1]. These countries were selected on the basis of the following criteria:
- Prior knowledge of interesting experiences with elements of rural water services delivery.
- Spread of development context and geographical regions, including very different general development indicators, such as their Gross Domestic Product (GDP) per capita, and levels of aid dependency (aid as percentage of Gross National Income).

Table 1: Study countries and some basic development sector indicators

Country	Income group[A]	GDP (US$/cap) (ppp)[B]	Official Development Aid as a percentage of Gross National Income in 2008[C]
Mozambique	Least developed country	934	21.6
Ethiopia	Least developed country	954	12.8
Uganda	Least developed country	1,196	11.7
Burkina Faso	Least developed country	1,304	12.6
Benin	Least developed country	1,445	9.6
Ghana	Other low income	1,551	7.9
India	Lower middle income	2,941	0.2
Honduras	Lower middle income	4,151	4.1
Sri Lanka	Lower middle income	4,769	1.8
Thailand	Lower middle income	8,060	- 0.2
Colombia	Lower middle income	8,936	0.4
South Africa	Upper middle income	10,244	0.4
USA	High income	46,381	N/A

Sources: [A] OECD, 2009 [B] IMF, 2010 [C] World Bank, 2010

- Capacity of IRC and its partners to carry out the study and follow-up activities.

From country case studies, three groupings can be identified (see Table 1):
- Least developed countries, which are highly aid dependent.
- Middle group of countries with mixed levels of aid dependency and income levels.
- Middle (and high) income countries, which are non-aid dependent.

The relative placement of a country across these different groupings can have an important influence on how the sector deals with the provision of rural water supply services. In the least developed countries, there is still a strong drive and need to rapidly increase coverage, whereas in the middle income countries the management and sustainability of services feature more prominently in sector agendas. The degree of aid dependency has a strong bearing on the need for, and importance of, aid alignment and harmonisation.

Other experiences with rural water services were included as part of the broader literature review, including those found in Latin America (Costa Rica, Nicaragua and Paraguay), South East Asia (Cambodia, Indonesia, Philippines and Viet Nam) and Africa (Senegal, Mali, Morocco and Niger). Countries in (post)-conflict situations or so-called 'fragile states' have been deliberately excluded from the selection process, as these pose very specific challenges for governance and sector capacity for the delivery of sustainable rural water supply that go beyond the scope of this study.

Methodology and common analytical framework

The methodologies for data collection followed a similar format in all study countries, employing a combination of secondary data collection, such as document and literature reviews, and consultation with key sector stakeholders through interviews and group meetings. Each study was conducted by a national expert, or team of experts, working as closely as possible with national authorities and non-governmental stakeholders.

In order to validate initial outputs of the country studies, and to gain sector buy-in to the results of the study, the majority of country studies incorporated a process of engagement with sector stakeholders in which preliminary findings were shared and discussed. This often involved a two-step process with those key issues identified at national level meetings being put to a group of experts and practitioners from district and regional levels who participated in similar workshops. This type of validation exercise served to enrich the conclusions in the studies, as well as jump-start a process of dissemination and dialogue.

The analytical tool

Provision of a water service, is often very context-specific. Culture, history, economy, politics, water resources, topography and demographic aspects all are determining factors in the possible levels of service, the opportunities to provide such a service, and the extent to which it can be financed sustainably. In short, what works in one place, does not necessarily work in another. And while this is a truism, we also know from long experience that there are a number of important elements that need to be in place to achieve more sustainable service delivery.

In order to provide a common point of reference for the various countries involved in this study, an analytical framework was developed, including a range of elements at three different institutional levels. This framework has drawn on earlier developed principle-based frameworks for rural water supply such as the one for scaling up rural water supply (Thematic Group for Scaling Up Community Management of Rural Water Supply, 2005), subsequently adapted by van Koppen, Moriarty and Boelee (2006) and van Koppen, et al., (2009), with a focus on multiple-use services.

The three main levels of analysis in the framework correspond firstly to the *national level* enabling environment; secondly to the *intermediate level* (most commonly the local or district government level or commune or municipality, depending on country context); and, thirdly, the *service provision level* (see Annex C). In total there are 18 elements included in this framework, each with a short description. It includes issues such as sector decentralisation and reform; institutional roles and responsibilities; financing; SDMs; learning and coordination; and monitoring and regulation.

The studies looked both at the formal policy and strategy documents for the respective countries, as well as the status of how it actually is on the ground. This often involved issues such as governance and political influence over (resource allocation) decision-making, and the relative strength of the rural water sector *vis-à-vis* other spending priorities. The expectation was that there would be a marked difference in the type of SDA followed between the least developed and most aid-dependent countries, and those that are at the middle to higher income level. The results of the individual country-level analyses were subsequently compared across all 13 countries in order to identify common trends or factors which seem to be important either as positive drivers of improved sustainability, or constraints to SDAs.

It should be noted that the focus of this study is on *rural* water supply. The definition of what is rural differs from country to country, and is often based on criteria such as population size (of settlements) or density. For this study, no new or common definition was developed. Rather, the criteria from the different countries where studies were carried out were respected. Some reference has been made to experiences with small town water supply, as the boundaries between small town and larger rural settlements are often blurred, or, as in Ghana and Uganda, small towns and rural growth centres are considered a particular sub-set of rural settlements. Small town water supply often presents a broader range of service provision options, some of which may be of relevance for rural water supply as well.

Rural sanitation has *not* been part of this study. Sanitation and water supply are, rightfully, often addressed in an integrated manner in implementation projects. However, fundamental differences exist between the two in terms of the type of service provided. The explicit omission of sanitation does not mean that that sanitation and water supply should be separated in service delivery. Rather, there is merit in studying how service delivery and sustainability mechanisms are different for the two specific interventions.

Key concepts and terminology

A number of important concepts are used in this book as part of making a comparative analysis across the country studies, and certain terminology is employed. The following sections explain the most significant of these concepts. Key definitions are also provided in Annex B: Glossary.

The Service Delivery Approach and Service Delivery Models

Accepting that there is a distinction between the physical system (the infrastructure) and the service which these systems deliver is a fundamental starting point. Service refers to the provision of a public benefit through a continuous and permanent flow of activities and resources; a concept applied

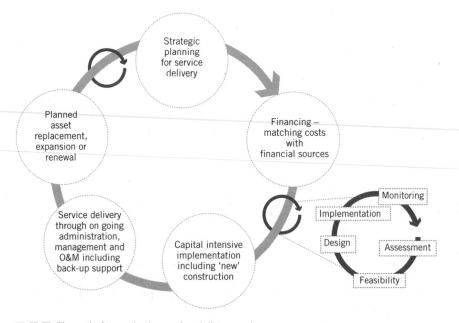

Figure 1: Stages in the service delivery cycle

in many other services, both in the developing and developed worlds, such as health, education, electricity, telephone and urban water supplies. A water service consists of access to a flow of water with certain characteristics (such as quantity, quality and continuity), as will be elaborated in the next section.

Water services delivery can be characterised by a life-cycle, consisting of various stages or phases over time (see Figure 1). A service normally starts with a capital-intensive period in which the physical systems are built and assets are developed at the front-end of the service period (i.e. the initial or 'new' construction of the physical system). These capital investments are typically done in the form of discrete implementation projects and often themselves follow a project cycle, with a series of consecutive activities including assessment, feasibility, design, implementation and monitoring. In the project cycle both physical construction as well as accompanying activities such as community mobilisation, training and other so-called 'software' activities are included.

This capital intensive period is followed by the stage of the service delivery itself, and consists of activities such as operations and maintenance (O&M) and administration, with the aim of the continued delivery of a service over time. Some of the activities include further physical interventions to update, expand and eventually replace physical assets (above ground structures, storage tanks, transmission pipes and pumps). Such activities happen in ongoing steps, but typically are again carried out through discrete projects. The cycle is closed when the service is expanded, e.g. by extending to households initially un-served, or by a major upgrade of the service, e.g. moving from a borehole with handpump to a borehole with motorised pump and small distribution system.

The concept of service delivery is not commonly used in rural water supply where most reference is made to the capital intensive or 'project' period only. **The Service Delivery Approach (SDA)** is a *conceptual* approach taken at sector level to the provision of rural water supply services, which emphasises the entire life-cycle of a service, consisting of both the hardware (engineering or construction elements) and software required to provide a certain level of access to water. We consider this to be the preferred universal approach, or paradigm, based on common principles for interventions at all levels from national to local, and one which should result in more sustainable services when compared to the more conventional infrastructure-focused approach which has been adopted under many projects and programmes in the past. The difference for sustainability between these two approaches is illustrated in Figure 2, where the left side shows the current reality for millions of rural people. Following construction of a new system (light grey rectangle), users have access to a given level of service (black line). The new system initially functions well, but due to lack of support and proper asset management quickly starts to deteriorate until it collapses completely, to be revived at some indeterminate time by the construction of a new system, typically by another agency.

The right side of the diagram shows the SDA concept: here, once a water system has been constructed, the service is maintained indefinitely through

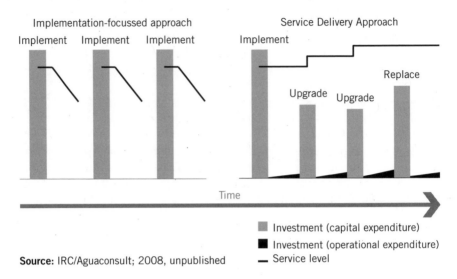

Figure 2: Water service delivery from the user perspective: repeated disappointment, or a Service Delivery Approach

a planned process of low intensity administration and management, with occasional capital-intensive projects for upgrading and eventual replacement.

When applied in practical terms in any given context, the principles behind the SDA are applied through one or more agreed **Service Delivery Models (SDMs)**, which reflect the realities of that country and the service area. SDMs provide agreed legal and institutional frameworks for delivering a service, including the levels of service (see next section) and commonly understood and accepted roles for public, private or community actors. In summary, the SDA provides the concept of a full life-cycle way of thinking, whereas the SDM describes the practical implementation of such an SDA.

Institutional levels and functions

In the study, reference is made to a number of different *institutional levels* within the conceptual framework for rural water service delivery. The definition of these levels is based on *functions* related to service delivery. Functions may or may not be linked to one or more specific institutional levels, depending on the degree of decentralisation and specific administrative hierarchy of the country. Broadly speaking, three distinct groups of functions can be identified with corresponding institutional levels:

- **Policy and normative functions – national level.** This refers to the overall enabling environment where sector policy, norms and regulatory frameworks are set, service levels defined, and macro-level financial planning and development partner coordination takes place. It can also be the level at which learning, piloting and innovation is funded and

promoted. Overall sector guidance and capacity building is set by this level of authority. This nearly exclusively takes place at national level, although in federal countries, states may also execute some of these functions.
- **Service authority functions – intermediate level.** This refers to the level where service authority functions such as planning, coordination, regulation and oversight, and technical assistance, take place. We use the term the intermediate level (i.e. in between the national and community level) of local government, such as district, commune, governorate or municipality, or whatever the exact administrative name given in a particular country, as a generic term to describe this level. In some cases the ownership of the physical assets of rural water supply systems is held by local government entities, but this varies from country to country. These functions may be split between different administrative levels, for example between provincial and district authorities, depending on the degree of decentralisation or mix between decentralisation and deconcentration of functions (see also section 'Decentralisation', p25).
- **Service provider functions – local level.** This refers to the level at which the service provider fulfils its functions of day-to-day management of a water service. This may also involve asset ownership (but this is rare) and investment functions under certain arrangements. Typically, the service provider functions are found at the level of a community or grouping of communities, depending on the size and scale of the water supply system(s) in question. The service provider function is fulfilled by a water committee under community management arrangements, or by an individual or business in other service provision options. This is the level at which day-to-day operation of the physical system takes place, and includes preventative and corrective maintenance, bookkeeping, tariff collection, etc. This may be done directly by a committee acting on behalf of the community, or in cases where there is professionalisation of community-based management, these tasks are increasingly delegated or sub-contracted to an individual (plumber or technician) or to a local company acting under a lease contract.

We argue that it is the interplay between these three levels that form an SDM. First of all, an SDM may include one or more management options (i.e. community-based management [CBM], private operator, self-supply, etc) for actual service provision at local level. These are guided by a country's existing policy and legal frameworks which define the norms and standards for rural water supply, the roles, rights and responsibilities, and financing mechanisms at national level. In the SDM, these management options should also then be supported by governance functions at intermediate level, including post-construction support, regulation, etc. Figure 3 provides an illustration of this. In a country, or even within a single decentralised or intermediate level administrative unit, there may be several SDMs, often related to the management options recognised in the national policy framework. Certain parts of

Enabling environment: policy, legal and institutional frameworks, macro-level investment planning, learning and innovation	National or state level
Service authority functions: planning, contracting, monitoring, post-construction support, learning	Local government
Service providers: day-to-day operation, administration and maintenance	System or community level

Community-based management	Public sector utilities	Private sector	Self-supply
Service Delivery Model	Service Delivery Model	Service Delivery Model	Service Delivery Model

Figure 3: Service Delivery Models

the SDM may be common between different management models. Certain service provision options may exist, but are not recognised, regulated or supported in a national framework. In such a case there is no full SDM. The SDA provides the conceptual framework within which these SDMs are defined.

Water supply service levels

A critical aspect of a service is the definition of the attributes of this service. There is of course a difference between accessing a borehole with a handpump which delivers 25 litres per capita per day (lpcd) to a person who may need to walk 500 metres, and having access to a continuous flow of water from a household tap. It is therefore important to define characteristics of a service a consumer has access to in terms of service attributes. The most commonly used service attributes are the quantity of water, its quality, the reliability and accessibility of supply, which is expressed typically as the distance between the water point and the homestead, or in terms of crowding (number of people with whom a water point is shared). A level is then a normative description of that service attribute. For example, access to 50 lpcd reflects a higher level of service than access to 25 lpcd. Some would argue that the costs or the affordability of the supply should be considered as part of the service level as well. While undoubtedly important, this is fundamentally different, as it is a reflection of the financial (or management) costs to get to a certain service level. It would often cost more to have access to 50 lpcd than to 25 lpcd. Costs are therefore not part of the service level definition itself, but reflect what is needed to reach a certain service level.

Various authors have tried to group some of these service characteristics together in the form of a 'service ladder' (van Koppen, et al., 2009; Moriarty, et al., 2010a). Figure 4, depicts the ladder developed by the WASHCost project (Moriarty, et al., 2010a), and shows a continuum running from 'no service' (which is effectively an insecure or unimproved source) to 'high-service', where

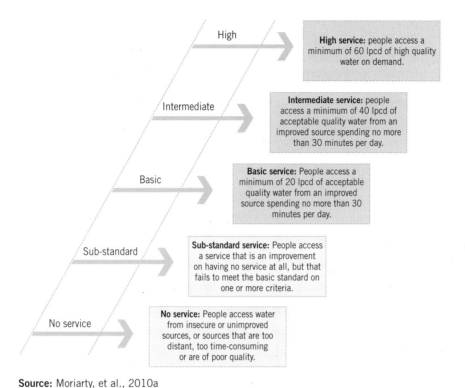

Source: Moriarty, et al., 2010a

Figure 4: Service levels

access is on demand at, or very close to, the household. There is often a reciprocal relation between the level of service provided and its sustainability. If a service is provided according to the expected level, consumers are often more willing to pay for a service. But, the costs of providing a higher level of service are normally also higher, and go beyond people's capacity to pay. Sustainability problems with a water service often reflect themselves through a slippage of the service level. With the exception of handpumps or submersible pumps, waters supply systems usually do not break down completely. Rather, there may be a gradual reduction in the level of service provided, going down the service ladder.

Sustainability

The concept of 'sustainability' is used liberally in the sector, and there are numerous interpretations of what this may mean in a wide variety of literature. In the more specific context of the rural water sector, many organisations define sustainability as the maintenance of the perceived benefit of investment projects (including convenience, time savings, livelihoods or health improvements) after the end of the active period of implementation. Hence, this definition may be closer to one that simply describes sustainability as: 'whether or not something continues to work over time' (Abrams, L.; Palmer, I., and Hart, T. 1998); meaning, in this case, whether or not water continues to flow over time.

Sustainability of the service is affected by a range of factors. These factors include not only the technical or physical attributes of the system, but also the financial, organisational (support functions) and managerial capacities of the service provider, which indicate the likelihood of the service continuing to be provided over time. Even though, in practice, different countries use (proxy) definitions and indicators for sustainability (which will be discussed in more detail in 'The status of access to sustainable rural water supply services', pxx), for this study we understand sustainability to be the indefinite provision of a water service with certain agreed characteristics over time.

Scale and scaling up

By the end of the 1990s many insights had been obtained into the factors affecting sustainability at community level, and evidence started to emerge of individual communities providing more or less sustainable services. However, the intervention cycles which led to these successes were often highly time- and resource-intensive or required significant amounts of post-construction support. Hence, an interest arose in the concept of 'scaling up', i.e. the application of community management approaches that would allow reaching a certain scale in sustainable service provision. The concept of scale and scaling up was further elaborated in various documents (e.g. Moriarty and Schouten, 2003; Lockwood, 2004). These highlighted the need to move away from a focus on the community level only, and towards one of working with organisations of the intermediate level, most importantly local government. It was argued that only at that level can functions such as planning, financing and post-construction support be organised that are needed to achieve sustainability at scale (i.e. working at an economy of scale). These, in turn, need to be embedded in national policies, regulations and institutional frameworks. In this book it is the notion of institutional arrangements that have the potential to achieve sustainability at larger geographical scale, well beyond a limited number of communities, that is encompassed by the term 'sustainable services at scale'.

For the term scaling up we follow the definition of Gundel, Hancock and Anderson, (2001) who distinguish between vertical and horizontal scaling up. Vertical scaling up refers to the institutionalisation of the functions and approaches that make sustainability possible; whereas horizontal scaling up refers to the application of these principles in a broader geographical area – what is also called 'scaling out' by some authors, such as Harrington, et al. (2001). Both institutionalisation and geographic spread are important to guarantee increased coverage and sustainability. In reality, the two processes are often difficult to separate, since geographical spread cannot take place without institutionalisation. In this book we use the term scaling up for the combination of the two processes.

Cost categories for the full life-cycle of rural water services

A clear and full understanding of all the costs required to deliver a reliable water service has been one of the major blind spots of the sector for decades (Fonseca, et al., 2011a and 2011b). Policy makers, governments and development partners of all descriptions have rarely, if ever, known the full costs of services, including not only the initial construction costs, but also what it costs to maintain them in the short and long-term including eventual replacement and expansion costs, as well as costs for both local and national level administration and planning. Without knowing what the real costs of an entire life-cycle of a service actually include, an informed debate about who should finance these costs is difficult if not impossible.

In both the individual country studies and in this book, the cost categories developed by the WASHCost project[2] have been used and are summarised in Table 2.

Decentralisation

Decentralisation is a core theme in many of the country studies, and is often a process that takes many years or even decades to reach a level of maturity in which lower tiers of government are not only given a mandate to deliver services, but are provided with adequate resources, capacities and decision-making power. Decentralisation has many meanings, but for the purposes of this study it can best be captured as 'the transfer of authority and responsibility for governance and public service delivery from a higher to a lower level of government', and the definitions are used based on the World Bank's Independent Evaluation Group definitions (World Bank/IEG, 2008).

In reality, and as supported by the findings of this study, there can be a number of pathways leading to decentralisation. These range from well planned and resourced processes that take place over many years, with progress indicators, to the so called 'big bang' decentralisation wherein the central level of government announces decentralisation, swiftly passes laws, and transfers responsibilities, authority, and/or staff to sub-national or local governments in rapid succession and without adequate time to embed sufficient capacity. The various aspects, or dimensions of decentralisation are set out in the left-hand column of Table 3; these are typically comprised of the transfer of administrative decision-making, power over financial control, and political or decision-making authority from central to lower levels of government.

Table 2: Definitions of cost categories

Cost category	Description
Capital expenditures – hardware and software (CapEx)	The capital invested in constructing fixed assets such as concrete structures, pumps and pipes. Investments in fixed assets are occasional and 'lumpy', and include the costs of initial construction and system extension, enhancement and augmentation. CapEx software includes once-off work with stakeholders prior to construction or implementation, extension, enhancement and augmentation (such as costs of once-off capacity building).
Cost of capital (CoC)	The cost of capital is the cost of financing a programme or project, taking into account loan repayments and the cost of tying up capital. In the case of private sector investment the cost of capital will include an element distributed as dividends.
Operating and minor maintenance expenditures (OpEx)	Expenditure on labour, fuel, chemicals, materials, regular purchases of any bulk water. Most cost estimates assume OpEx runs at between 5% and 20% of capital investments. Minor maintenance is routine maintenance needed to keep systems running at peak performance, but does not include major repairs.
Capital maintenance expenditure (CapManEx)	Expenditure on asset renewal, replacement and rehabilitation costs, based upon serviceability and risk criteria. CapManEx covers the work that goes beyond routine maintenance to repair and replace equipment in order to keep systems running. Accounting rules may guide or govern what is included under capital maintenance, and the extent to which broad equivalence is achieved between charges for depreciation and expenditure on capital maintenance. Capital maintenance expenditures and potential revenue streams to pay those costs are critical to avoid the failures represented by haphazard system rehabilitation.
Expenditure on direct support (ExpDS)	This includes expenditure on post-construction support activities direct to local level stakeholders, users or user groups. In utility management, expenditure on direct support such as overheads is usually included in OpEx. However, these costs are rarely included in rural water and sanitation estimates. The costs of ensuring that local government staff have the capacities and resources to help communities when systems break down or to monitor private sector performance are usually overlooked.
Expenditure on indirect support costs (ExpIDS)	This includes macro-level support, planning and policy making that contributes to the service environment, but is not particular to any programme or project. Indirect support costs include government macro-level planning and policy-making, developing and maintaining frameworks and institutional arrangements, and capacity building for professionals and technicians.

Source: Fonseca, et al., 2010

Table 3: Dimensions and modes of decentralisation

Dimensions of decentralisation	Modes of decentralisation
Administrative decentralisation – how responsibilities and authorities for policies and decisions are shared between levels of government, and how these are turned into allocative outcomes	**Deconcentration** – the shallowest form of decentralisation, in which responsibilities are transferred to an administrative unit of the central government, usually a field, regional, or municipal office
Fiscal decentralisation – the assignment of expenditures, revenues (transfers and/or revenue-raising authority), and borrowing among different levels of governments	**Delegation** – in which some authority and responsibilities are transferred, but with a principal – agent relationship between the central and lower levels of government, with the agent remaining accountable to the principal
Political decentralisation – how the voice of citizens is integrated into policy decisions, and how civil society can hold authorities and officials accountable at different levels of government	**Devolution** – the deepest form of decentralisation, in which a government devolves responsibility, authority, and accountability to lower levels, with some degree of political autonomy

Source: World Bank/IEG, 2008

Aid effectiveness

Given that the WASH sector is still highly aid-dependent in many countries, aid effectiveness has for some time been a concern for both the donor community and recipient governments working in the sector. Figure 5 provides a model of the main components and terminology regarding harmonisation in the context of aid effectiveness.

The end goal of achieving aid effectiveness is country *ownership*, which means that the political agenda is driven by the needs and priorities of the recipient (partner) country, rather than by development partners. Thus, recipient countries are encouraged to develop the necessary policies, strategies, programmes and public financial management systems through which the aid can be channelled. A second supporting element is then *alignment*, in which development partners align their aid to the country partner's agenda as well as the country's systems, such as their financial and monitoring systems. Within this approach donors come together to *harmonise* their efforts so that common arrangements are established, procedures simplified, and information shared.

With a common agenda and national sector programme, plus better alignment and harmonisation, the development effort can be managed for results (and impact) rather than being managed on a project by project basis where it is difficult to determine the overall results. Where alignment and harmonisation take place, there is the opportunity for mutual accountability, with development partners being accountable to support the partner in funding the agenda (sector programme), and the partner country being accountable for what is achieved with the development partner aid.

28 SUPPORTING RURAL WATER SUPPLY

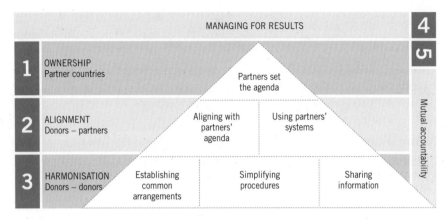

Source: OECD, 2006

Figure 5: Main components and terminology related to aid effectiveness

The terminology and concepts mentioned above apply equally at national and intermediate level. In reality, one may find a discrepancy between levels, for example, where there is a Sector Wide Approach (SWAp) at national level, but where some donors (including international non-governmental organisations [INGOs]) still have parallel projects at intermediate level.

Moving towards the application of one or more agreed SDMs requires not only harmonisation between development partners and government, and between development partners, but also across different government agencies, between national and intermediate levels of government, and between other sector stakeholders such as water resources authorities, water service providers and communities. In this context we understand *coordination* to refer to the mechanisms (both formal and informal) through which these actors articulate their activities and strategies amongst each other, and how they negotiate their role in or contribution to the sector.

Endnote

1 Where reference to the studies is made in the text, the specific authors are not cited, but the full references are made available in the References section at the end of this book.
2 WASHCost is a 'sister' project to Triple-S, also managed by IRC. It carries out research into methods for collecting and collating information relating to the real dissaggregated costs in the life-cycle of water, sanitation and hygiene service delivery to poor people in rural and peri-urban areas. For more information see: http://www.washcost.info/page/121.

CHAPTER 3
Country sketches

This study deliberately includes a range of countries at differing points on the long-term development of their respective rural water sectors. The summaries in the following pages are intended to provide a brief sketch, or overview of each of these as context for the subsequent analysis and may be useful to read as a 'snapshot' of where the sector is in 2010. Each summary includes basic information on the country, the administrative set up and explains the most important institutions and agencies involved in rural water provision. It then highlights the particular SDMs that have been established in each country and closes with a brief assessment of the key issues facing the sector.

Benin

Benin is located in West Africa and has a population of approximately 8.8 million. The economy of Benin is largely dependent on subsistence agriculture, cotton production and regional trade. Benin is classified as a least developed country with a GDP per capita of US$1,445 (IMF, 2010). At the end of the 1980s, Benin underwent a major political crisis. A massive reorganisation of the State and a decentralisation process resulted from this crisis.

Administrative set up

Benin is divided into 12 departments, and subdivided into 77 communes.

Main sector institutions

- The **General Directorate of Water** (DG Eau) is the main sector agency, responsible for policy development, facilitation and regulation. It has deconcentrated offices at departmental level.
- The Société Nationale des Eaux du Bénin (SONEB) is the National Water Utility of Benin, and is the main service provider in cities and towns.
- **Communes** are the level of local government which, since early 2003 when the process of decentralisation began in the water sector, has the obligation to guarantee water supply service delivery. Among others, communes are responsible for investing in facilities, and they have become asset holders responsible for ensuring that service delivery arrangements are in place. In reality, the decentralisation process is going very slowly, and many communes do not play their role as asset holders. Many planning functions continue at departmental level.
- **Operators** can be either private operators or community groups.

The sector has been largely guided by the **Rural Water Supply Assistance and Development Programme** (PADEAR), started in the 1990s. This programme has been aimed at reorganising the sector, as well as at harmonising efforts in rural water supply among sector stakeholders.

Service Delivery Models

The national water supply strategy was published in 2006, and is aimed at strengthening the decentralisation process. It identifies the need to promote delegated management as a way of improving sustainability of supplies. The specific delegation model depends on the type of infrastructure, either small piped systems or point-source infrastructure, and the already existing management arrangements. As a result, in rural and small town settings, many water SDMs coexist:

- In the more complex distribution systems, four models of delegation have been identified in the national strategy using lease contracts: to a private operator; tripartite involving the communes, water user association and private operator; production by a private operator and distribution by a water user association; and to a water user association. Under these models, the commune owns the infrastructure and undertakes planning, the district directorate for water provides support and ensures regulation, and the operator is the one who provides the services, with different responsibilities regarding asset management, depending on the type of delegation. The operator is either private or community-based, but the former is strongly encouraged by the strategy.
- In the case of point-source infrastructure, four delegation models also exist: to a community representative, a private operator, several systems to one local operator, or to an operator of both point-sources and complex systems in a geographical area. The commune delegates following a proposal for a community representative, or to a private operator following a feasibility study of the financial sustainability of the system and subsequent tender.

Key issues

Although, the actual implementation of these models lags behind, there are notable changes in professionalisation of community management. The study identified a number of interesting case studies with variations of the above-mentioned models. The Benin study highlights the potential that delegated management models has for professionalisation of service provision.

Burkina Faso

Burkina Faso is a relatively small country in West Africa with a population of approximately 14 million, of which 77% live in rural areas. The decentralisation process started in 1998 with the adoption of several decentralisation laws and decrees. Despite positive growth rates in the past years, a significant number of people – up to 46% in 2004 – remain below the poverty line. It has a per capita GDP of US$1,304 (IMF, 2010). Current coverage for rural and peri-urban water supply stands at around 52%, but there are significant disparities between regions.

Administrative set up

The country is divided into 13 regions, 45 provinces, 351 communes and about 8,300 villages.

Main sector institutions

Though the ultimate responsibilities over a number of small towns and the villages located within the boundaries of urban communes are unclear, sector organisation is relatively well defined, with a separation of functions between a number of key institutions:

- The **Ministry of Agriculture, Fisheries and Water Resources** is the ultimate authority for water supply and sanitation issues in rural areas, and the **National Office for Water and Sanitation** (ONEA), is responsible for urban areas.
- The **General Directorate for Water Resources** is the national body for policy development and planning.
- The **Regional Directorates for Agriculture, Fisheries and Water Resources** coordinate planning, resources and activities at regional level.
- The **National Office for Water and Sanitation** and the **private sector** operate some dimensions of water services in a selected number of communes.
- **International donors** (GIZ [German international cooperation organisation], World Bank, the *Agence Française de Développement* [AFD – French Development Agency], Danish International Development Assistance [Danida], the European Union [EU], etc.) finance about 90% of the national water budget.
- **INGOs** implement and support local capacities.
- **Communes** are responsible for ensuring service provision to their populations.

Service Delivery Models

In Burkina Faso there are four principal SDMs recognised in sector policy, with three of them being recent and under trial in specific regions (2–4):
1. Community-based management, covering 70% of rural and peri-urban water supply systems.
2. Delegation to the private sector in the form of leasing contracts. Three major operators share the market.
3. Delegation to ONEA through mutual agreement with urban communes for water provision in their urban perimeter for a specific length of time. A local non-governmental organisation (NGO) supports the commune in negotiations with the Agency. Seven communes have adopted this model.
4. Through Associations for Water Conveyance Development. These Associations aim to accelerate growth of the water market in rural and semi-urban areas through public-private partnerships (PPPs) extended to associations from civil society and NGOs. The service delivery relies on a voluntary pooling of equipment.

Key issues

CBM was implemented thoroughly prior to the 2000 National Water Reform, which supports the implementation of PPPs in the management of water delivery services. Such partnerships are recognised through leasing contracts (for delegation) or through coordination units (for pooling equipment). Such experiences are however still conducted at pilot level, and no scaling is planned for by the six year National Water and Sanitation Program (2009–2015) to meet MDGs. Moreover, a number of local committees created before 2000 are hostile to change, especially those who still keep their systems going, and want to keep their leadership in public services delivery. In parallel, the central government created an executing agency to operate water services in place of the communes, with the objective of accelerating the pace to meet the MDGs. By doing so, the government holds back the transfer of project ownership to local authorities. Hence, the central government is faced with both the need to produce immediate results and the need to set up ownership at communal level.

A certain level of coordination is in place between national stakeholders, donors and actors (NGOs, General Directorate for Water Resources, etc.), notably through the National Water and Sanitation Program and its numerous platforms. The main issue is around the actual transfer of responsibility and resources to the communes whose capacities to plan and manage water services are limited.

Colombia

Colombia has around 45 million inhabitants, 76% of whom live in urban areas. It is a lower middle income country with a GDP per capita of around US$8,936 (IMF, 2010). In spite of high levels of economic growth, huge inequalities exist; it is among the top 10 countries with the highest Gini-coefficient in the world. Steady but slow progress is being made in reducing poverty levels; but a significant percentage of the population continues to live in poverty – mostly in rural areas. The process of decentralisation began in the 1990s.

Administrative set up

Between the national and municipal levels of government there are 32 departments. There are approximately 1,100 municipalities (the lowest level of local government).

Main sector institutions

The sector has been guided by the Law 142 on Water Supply and Sanitation Services Provision since 1992. It clearly separates roles and responsibilities.

- The **Ministry of Environment, Housing and Territorial Development** (MAVDT) is the ministry responsible for water supply. Policy development and coordination are amongst its tasks.
- The **Water Regulatory Commission** (CRA) and the **Public Domestic Service Superintendent** (SSPD) form the two key regulatory bodies. The CRA is responsible for setting the regulations, whereas the SSPD has a monitoring and enforcement role.
- **Departments** historically have had a limited role in water supply. In the last administration however, they were given an important role in planning and financing investments in new infrastructure and rehabilitation.
- **Municipalities** are constitutionally responsible for guaranteeing access to water and sanitation services. This implies establishing service provision arrangements. In addition, they are responsible for planning and financing investments. Finally, they have a role of providing technical assistance to service providers.
- Four different types of **service providers** are recognised by the legal framework: direct provision by the municipality through a municipal company; private providers; mixed public-private companies; and, community-based service providers (which are further sub-divided in four modalities).

Service Delivery Models

CBM has been firmly established as the main Service Delivery Model in rural areas and even in many small towns. This finds its roots in a long tradition of CBM in the country, but supported by the Law 142. It is estimated that there are over 11,000 community-based water supply and sanitation service providers. Though other SDMs exist as mentioned above, these are mainly applied in towns and cities.

Key issues

Since the second half of the 1990s it has also been recognised that many community-based service providers suffer from technical and administrative deficiencies, and a range of models of support have come into existence to professionalise CBM. These can be classified as:
- Direct support to service providers by the MAVDT, through a programme called 'Business Culture Programme'.
- Support by departmental authorities as part of larger infrastructure development and rehabilitation programmes.
- Support by municipalities to community-based service providers. In some cases it has been municipal staff providing post-construction support directly; in other cases, the municipality contracts the urban utility to support rural operators in the neighbouring areas.
- Technical assistance through civil society or private sector agencies. One example of a civil society initiative is the Asociación Colombiana de Organizaciones Comunitarias Prestadoras de Servicios de Agua y Saneamiento (AQUACOL) (Colombian Association of Community-Based Water and Sanitation Services Providers). It facilitates mutual support between its members. An example of the private sector is the Coffee Growers' Association which supports the technical and administrative management in villages where it has helped to build water systems, predominantly in the departments where coffee growing is concentrated.

Ethiopia

Ethiopia is Africa's second most populous country and, while it is among one of the world's poorest, with a per capita GDP of US$954 (IMF, 2010) the government has committed itself to very ambitious targets to expand access to water and sanitation across the country. These targets are generally supported by a technologically-driven strategy of constructing new water supply systems in rural areas.

A review of the **Ministry of Water Resources** (MoWR) Accelerated Universal Access Plan – a key sector policy document – suggests that the country's target to increase coverage of water access (defined as 15 lpcd potable water within 1.5 km) to 98% by 2012 is more ambitious than the MDG target. Decentralisation has been ongoing since 1991.

Administrative set up

It is a federal state with nine regions (and two chartered cities) and approximately 550 *woredas* (districts).

Main sector institutions

- The **Ministry of Water and Energy** (MoWE), at the federal level, is responsible for preparing national water policy, strategy and action plans, and for establishing national standards for, amongst other things, water quality and infrastructure. The ministry is also accountable for overseeing the implementation of policy and strategy. In addition to this regulatory role, the ministry gives technical advice (in the form of manuals and guidelines) to **Regional Water Bureaus**. The MoWR also manages the implementation of the largest capital investment projects.
- At the regional level the **Bureau of Water Resource Development** is responsible for the implementation of federal policies by adapting them to the specific conditions of the region. The Water Bureaus also have a regulatory role for certain tasks as delegated to them by the ministry.
- **Zonal Water Resources Development Offices** support the Water Bureaus in giving technical support to *Woreda* **Water Offices** and **Town Water Supply Offices**. They are accountable for coordinating activities, plans and reports from *woredas*, and liaising between regional Water Bureaus and *Woreda* Water Offices.
- *Woreda* **Water Resources Development Offices** are responsible for the investigation, design and implementation of small-scale water supply schemes. In towns where there are no municipalities, they are also responsible for providing technical support to the Town Water Supply Offices.
- The *Woreda* Water Offices each have a *Woreda* **Water Supply Sanitation and Hygiene Team** made up from the offices of health, education,

women, and agriculture. These teams are responsible for planning and implementation of water and sanitation activities.
- O&M under CBM is the responsibility of the **users** and, more specifically, the **WASH Committees** (WASHCO) (or in towns, the **Water Boards**). WASHCOs are responsible for making minor repairs for which they charge a tariff to users. WASHCO members are selected from the communities through an election process.

Service Delivery Models

The formally recognised SDMs are:
1. Community-management is the main SDM implemented in the rural water sector. After construction (mainly supported by donor funded projects and NGOs) and the handover of schemes, operation and minor repairs are handled by the WASHCOs representing the community. In multi-village schemes, Water Boards are established to oversee these tasks. Water Boards comprise representatives from the WASHCOs of individual villages.
2. Self-supply is a low-cost approach to service delivery initiated by individual families or groups. Within this model, water sources, usually hand-dug wells, are constructed with limited direct support. Support to establish facilitating markets and supply chains is required to scale up. In 2009, this low-cost approach was formally recognised in policy. However, self-supply has yet to be incorporated as a formally recognised model in sector performance assessments.
3. Municipalities with Town Water Boards in small towns. Small town water supply systems are managed by Water Boards, usually with support from and reporting to municipalities.

Key issues

A sector harmonisation programme (known as One WASH), which started in 2006, has widespread commitment and is gathering momentum. There is a strong drive from development partners and donors at national level towards harmonisation (particularly on funding mechanisms). Emphasis is on construction and implementation, which is done relatively well, but with much less focus on providing a long-term service.

Ghana

Ghana has an economic growth rate, Human Development Index (HDI) and corruption indices that far-outstrip its neighbours, and a GDP per capita of US$1,551 (IMF, 2010). Just over half of the population lives in rural areas. In the last five to ten years the focus in rural water supply has shifted from point sources towards simple piped networks for small towns, with a reported average coverage rate in 2009 of 57% in rural areas. However, according to 2008 Demographic and Health Survey (GSS, GHS and ICF Macro, 2009), the percentage of the rural population with sustainable access to an improved water source was 76.6%. Since the late 1990s, Ghana has implemented comprehensive local government and decentralisation reforms.

Administrative set up

Ghana has a four-tier structure: national, regional, district and sub-district. There are 10 administrative regions, which are divided into 170 Metropolitan, Municipal, and District Assemblies (MMDAs).

Main sector institutions

- Rural water supply is the responsibility of the **Ministry of Water Resources Works and Housing** (MWRWH). The ministry has the primary responsibility for the formulation of policies for the water sector.
- The **Water Directorate**, established in 2004 as a division within the MWRWH, is expected to coordinate the activities of all key sector institutions operating under the auspices of MWRWH.
- The **Ministry of Local Government and Rural Development** (MLGRD) is the main actor responsible for overseeing local government in the form of MMDAs.
- The **Community Water and Sanitation Agency** (CWSA) is responsible for rural water: namely, water supply to scattered rural communities and small towns, while the **Ghana Water Company Ltd** and **Aqua Vitens Rand Limited** (AVRL) are responsible for urban water supply.
- **Regional Coordinating Councils** (RCCs) have the mandate to monitor, coordinate and evaluate the performance of all MMDAs.
- **MMDAs** exercise deliberative, legislative and executive functions, and are responsible for the overall development of the districts. Water is not expressly noted among the functions of the MMDAs, and it may be one reason why it falls low on the list of priorities.
- Within every District Assembly (DA) there is a **District Water and Sanitation Team** (DWST) which is a technical unit to support the delivery of water and sanitation services. In small town contexts the DA normally delegates responsibility to **Water and Sanitation**

Development Boards (WSDBs) to manage and hold the water systems in trust.
- **Water and Sanitation Committees** (WATSAN) are set up around point sources, such as a handpump. They set water user fees (with approval from the DA), maintain accounts, and manage day-to-day operations of these water points.

Service Delivery Models

There are four broad groups of SDMs:
1. CBM – there are a number of different types in operation, dependent on population size and technology, and employed mainly in rural and small town contexts.
2. Utility managed, including PPPs, with a management contract and community partnerships with a utility for bulk supply.
3. Private providers, including a broad group of largely unofficial models that have emerged more or less spontaneously to meet the demand for services not met by the two official models.
4. Self-supply which has evolved as a response to the inadequate formal water service delivery systems. It is, however, not addressed in policy or strategy papers.

Key issues

Sector support remains almost entirely a bilateral affair between sector agencies and development partners and, while most players express strong verbal support for harmonisation, progress appears to be mixed, with only an *ad-hoc* Sector Working Group which essentially serves as a platform for information sharing between government and donors. A critical issue in pushing the harmonisation agenda has been the past level of government commitment. However, there now appears to be a genuine desire on behalf of government and at least some donors to move towards a more harmonised approach. A SWAp roadmap was established in 2009 to build towards sector-wide planning and coordination.

Honduras

Honduras is a relatively small country with a population of some 7.5 million, of which approximately 4.2 million live in rural areas. Economic growth rates have been high in the past years, yet a significant number of people – up to 50% – remain below the poverty line. It has a GDP per capita of US$4,151 (IMF, 2010). Decentralisation started in 1990 with the Municipal Law. Current coverage for rural water supply stands at around 77%, but it is significantly lower for those living in highly dispersed small rural communities. The political situation has been highly fragile in recent years.

Administrative set up

The country is divided administratively into 18 departments, and there are 298 municipalities, with clear definitions between rural and urban areas.

Main sector institutions

- In common with most countries in the Latin American countries region, the **Ministry of Health** is the ultimate authority for water supply issues. Sector organisation is relatively well defined, with a separation of functions between a number of key institutions:
- **Consejo Nacional de Agua Potable y Saneamiento** (CONASA) is the National Council for Water and Sanitation, and is the apex body for policy development and planning.
- **Ente Regulador de los Servicios de Agua Potable y Saneamiento** (ERSAPS) is the sector regulatory body responsible for setting normative standards and performance criteria for operators.
- **Servicio Autónomo Nacional de Acueductos y Alcantarillados** (SANAA) is the National Autonomous Service for Water and Sewerage. It is a parastatal that was previously responsible for direct implementation and support functions, but under the new reform framework it provides support, and direct implementation should fall to municipalities.
- **Municipal** and **urban authorities** are responsible for ensuring service provision to their populations.
- **National and international NGOs** play a significant role in both the construction of new systems and support to O&M, accounting for some 15-20% of implementation, but generally with a semi-permanent presence.

In addition there is the Fondo Hondureño de Inversión Social (FHIS) (Honduran Social Investment Fund), which is the largest implementing programme in the country accounting for over half of all direct investment in the water sector.

Service Delivery Models

There are four principal SDMs recognised in sector policy:
1. CBM, – although for larger communities (i.e. more than 500 households or with multi-village schemes) voluntary models are considered to be limited and more professionalised approaches are the norm.
2. Municipal provision through small-scale utilities run by local government.
3. Utility provision – public or private.
4. Self-supply, which refers to highly dispersed small rural committees (of up to about 250 people) formally recognised, but with no formal support programmes).

Key issues

CBM has been heavily promoted and generally well implemented in the past. In addition there is a well-documented track record in the area of post-construction support and sector information systems, producing several innovative models over the last decades. As a result, sustainability of rural water supply has been relatively well-managed, particularly with support from development partners (United States Agency for International Development [USAID] and large INGOs). But in recent years this support has reduced significantly with the associated deterioration of support systems.

More recent changes under the sector modernisation process, started in 2003, but largely stalled over the past five years, are promising and include many of the elements and institutions that are required for adopting a sustainable SDA on paper (coordination and planning platforms, a regulator, legal recognition of CBM options, monitoring systems, etc,). However, in practice, these elements are not adequately funded and, most importantly, are not co-ordinated in a coherent way. The sector remains largely donor dependent, but determining the exact level and sources of funding streams is challenging as there is no single annualised overview of financial flows or clear financing policy. There are also limited consolidated planning and investment processes. The result is a somewhat fragmented sector which lacks the final 'kilometre' of finalising policy and improving national coordination mechanisms.

India

Although it has a GDP per capita of US$2,941 (IMF, 2010), some 42% of India's 1.2 billion people still survive on less than US$1.25 a day (World Bank, 2008). JMP reports show an increase in the population with access to improved sources of rural water supply from 58% in 1990 to 73% in 2008 (WHO/UNICEF, 2010). However, in the last decade the problem of 'slippage' has been highlighted, and government statistics put rural coverage at only 67% in 2009.

Administrative set up

The country has a three-tiered federal democracy with central, state and district governments. There are 28 states and 7 union territories, and 631 districts.

Main sector institutions

- Government responsibility for rural water at national level is shared by three major ministries, the **Ministry of Water Resources** that looks after irrigation and river waters, the **Ministry of Environment and Forests**, amongst others responsible for water pollution, and the **Ministry of Rural Development** that oversees watershed development, as well as rural drinking water supply and sanitation through the **Department for Drinking Water and Sanitation** (DDWS). The DDWS is the apex agency for rural water nationally, and is responsible for policy formulation, preparing federal budgets, setting norms and providing the bulk of the funding for rural water supply provision.
- Under the federal system, it is the individual **states** that are ultimately responsible for rural water provision. At state level there is a similar division of responsibility as at national level. **Departments for Rural Water Supply** or **Public Health Engineering** are implementing rural water supply and sanitation programmes. Some states have different arrangements for rural water supply, with dedicated **Water Boards** responsible for bulk water supply to rural communities.
- Responsibility for rural water provision lies at the lowest tier of government, the *Gram Panchayats*, but the decentralisation process has been varied and, in many areas, remains incomplete. Under the auspices of the *Gram Panchayats*, community-based organisations (CBOs) called *Pani Samitis* (Village and Water Supply Committees) are the main institution responsible for community provision; in some cases such CBOs are officially incorporated as sub-committees of the *Gram Panchayat*. In progressive states like Gujarat, Maharashtra and Kerala, CBOs out-source construction and O&M of water supply systems to other CBOs, NGOs and private sector companies.
- Since almost all investments in India are supported by either federal or state funding, assets remain the property of the state, with the right to manage systems delegated to the *Gram Panchayats*. There is no formal regulator of the rural water sector in India, although one state, Maharashtra, has established a **Water Resources Regulatory Authority**.

Service Delivery Models

Since 2002, India has had a formally recognised national community-based rural water supply programme, which has been translated into various state level models over the years. In some models, state sector technical agencies have responsibility for construction and handing over to community management entities (e.g. the *Jalswarajya* model from Maharashtra). In others, community entities enjoy full financial autonomy for the planning, design and oversight of contractors to build systems and then take over day-to-day management and O&M functions (e.g. the Water and Sanitation Management Organisation [WASMO] model from Gujarat).

Key issues

The issue of slippage, or non-functionality, has become a high priority concern of the DDWS in recent years, culminating in a recent plan by central government to monitor targets for a range of indicators including coverage, source protection, tackling of fluoride and arsenic, and poorly served Scheduled Castes. There is also a growing concern over water security in many states – due to changes in rainfall patterns, contamination from industrial and agricultural sectors, and competition for water use from various sectors, including urban consumers. Water security planning for rural populations focuses on multiple sources to guarantee supply, and has reemphasised the importance of traditional sources, such as rainwater, and of source protection measures. As coverage levels have increased so has the concern with water quality issues. There has, however, been a recent policy thrust towards improving rural water supply as evidenced by the National Rural Drinking Water Quality Monitoring and Surveillance Programme of 2006, the National Drinking Rural Water Supply Programme of 2009, the Results Framework Document of 2010, and two key documents currently under preparation – the Strategy for Rural Drinking Water Supply and the Twelfth Five Year Plan (2012–2017).

Mozambique

Following the resettlement of civil conflict refugees and relative political stability, Mozambique saw consistently high economic growth rates, but these were from a very low base and extreme poverty is widespread, reflected in a GDP per capita of US$934 (IMF, 2010). The country has an estimated population of 23 million, of which 63% live in rural areas. The decentralisation process started through legislation from 1997 onwards, but has proceeded very slowly, and transfers of capacity, financing and real decision-making power are very limited. In the water sector there is still a very high de facto level of centralisation and power held by the ministry at central level, with some limited deconcentration to provincial level – for example, provinces now have more of a role in monitoring.

Administrative set up

There are 10 provinces comprising 128 districts.

Main sector institutions

- The **Ministry of Public Works and Housing** (MPWH) is the most important government institution for rural water at national level.
- The **Direcção Nacional de Águas** (DNA) (National Directorate for Water) sits within the MPWH, and is responsible for policy, strategy, norms and channelling financing to lower levels. The **Departamento de Água Rural** (DAR) (Department of Rural Water) is in turn housed within the DNA.
- At provincial level the **Provincial Department of Public Works and Housing** (DPOPH) has a role in supporting and compiling plans from districts and supporting and monitoring work of districts, but there is no presence of the DAR and therefore there is an institutional gap.
- At district level the **District Water and Sanitation** (DAS) is responsible for local planning, implementation and oversight of contracting.
- At village level **water committees** are responsible for ongoing operation and small-scale maintenance and minor works.

Service Delivery Models

Sector policy is based on Village Level Operation and Maintenance (VLOM) and the demand-responsive approach (DRA), with some more minor modifications in provinces to do more to support O&M. But essentially there is only one formally recognised SDM which is CBM with water point committees; no other formally recognised models for rural areas exist, and there are very few piped systems in defined rural areas. There is no formal system or institutional mandate for post-construction support. In theory, water provision is demand driven, but there is very limited capacity for follow-up either at district or provincial level, and no funds are made available for direct support. Post-construction support, especially in the northern half of the country, is complicated by huge distances, very low population densities, and very limited commercial capacity. As a result, there is a general divide between the north and south where in the south CBM/VLOM is able to function (with some problems), and in the north it is extremely challenging.

To date, self-supply has not been recognised as an acceptable solution for rural water. The legal status of water committees is not clear, but there are doubts as to whether they would be recognised as a legal entities.

Key issues

The rural sector in Mozambique can be characterised as being very much focused on implementation of new systems as a natural drive to increase coverage from a very low base, a process which in the past has been driven by often uncoordinated donor funded programmes. A rural SWAp is now under preparation which will establish a coordinating secretariat to set up platforms at national and provincial level. Most donors already communicate with each other through a well-established roundtable and troika leadership model. This initiative for the rural sector is part of a broader effort by the Mozambican government and its partners, which signed a Code of Conduct agreement in 2008.

South Africa

South Africa is an upper middle income country with an abundant supply of resources, and a modern infrastructure. The country accounts for almost 45% of the GDP of the entire African continent. Over the past ten years the economy averaged a growth rate of just under 3% per annum, however, in the past few years, this has grown to almost 5% a year. It has a GDP per capita of US$10,244 (IMF, 2010), placing it as a middle income country; however, there are high levels of poverty, and South Africa has a Gini-coefficient of 57.8. The country has an estimated population of just under 50 million, with around half living in small towns and rural areas. In terms of coverage some 91% of the population currently have access to improved water supply; almost all of the unserved reside in rural areas, where 22% of the population still does not have access to an improved water supply. South Africa has undertaken a comprehensive decentralisation process, which started in the early 2000s, with a well elaborated framework and a clearly defined water service authority mandate at local government level.

Administrative set up

There are nine provinces with six metropolitan and 46 district municipalities which are further divided into 231 local municipalities.

Main sector institutions

There is a well defined division of roles and responsibilities for the delivery of WASH services in South Africa, with the national government having responsibility for setting norms and policies, as well as a major support and leadership role to ensure a strong and collaborative sector. Local government is responsible for ensuring the actual provision of services and universal access to services:
- The national government is represented by the **Department of Water Affairs** (DWA), as a policy, norms and standards setter; provincial (deconcentrated) offices of DWA play a technical capacity role through regional offices which also provide support to local government.
- Within local government, **water services authorities** (WSAs) have been established to cover water policy for free basic services. They have a technical department responsible for infrastructure such as water, roads, etc.
- **Water services providers** (WSPs) at system or multi-system or municipal level provide day-to-day O&M, customer care, revenue collection, etc.

Service Delivery Models

There is a range of WSP institutional arrangements, from options that cover an entire district municipal area (or even multiple municipal areas) to individual CBO options that cover specific communities. WSAs decide on the provider through contracted arrangements. Three principal SDMs can be identified in sector policy:
1. Municipal provision: this can be either directly by the municipality itself through a service unit, or via contracting out to a municipal-owned utility.
2. CBOs contracted as service providers by the WSA: they often receive support from (often private) Service Support Agents (SSAs) which both provide technical back-stopping and a monitoring and reporting function to the WSA.
3. Private sector companies: different types of contracts can be entered into with the private sector ranging from concessions to lease and management contracts.

Key issues

One of the strengths of South Africa is the approach which has established a range of frameworks to enable water services provision at the local level with service targets and financing, planning, regulation, monitoring, reporting and support. The institutional framework for water services also recognises that there is no single model or institutional arrangement that can address the different realities at the local level, and thus legislation requires a vigorous assessment process to find the most appropriate service provision arrangement(s). One of the other highlights is the SWAp (called *Masibambane* or 'working together') launched in 2001 which has fundamentally changed the way the water sector operates, by coordinating the mandates and relationships between different stakeholders towards a truly scaled-up effort. However, the rural water sector is not without significant problems in so far as it has in fact largely focused on rapidly increasing coverage through implementation and rehabilitation, with a resultant capital maintenance backlog; this, despite the fact that it potentially has the systems in place to start addressing the full life-cycle of services in a more structural way.

Sri Lanka

Sri Lanka is a lower middle income country with a population of some 20 million people, of which approximately 78% live in rural areas. Coverage of improved water supply nationally is high – a recent Asian Development Bank report puts the national average at 82% (Asian Development Bank, 2010). In rural areas coverage is estimated to be 72%, but there are wide discrepancies and much lower rates in the eastern and northern parts of the country.

Administrative set up

The governance structure has four levels: central government, Provincial Councils, *Pradeshiya Sabhas* and village-level organisations. The country is divided into nine provinces and 25 districts, with districts further sub-divided into 326 divisions. *Pradeshiya Sabhas* are the lowest unit of government, and their jurisdiction largely coincides with divisional boundaries. There are 270 *Pradeshiya Sabhas*.

Main sector institutions

- The **Ministry of Water Supply and Drainage** (MWSD) is the apex body responsible currently for urban and rural water supply in the country, while the **Ministry of Irrigation and Water Management** is responsible for regulation and control of inland water.
- The **National Water Supply and Drainage Board** (NWSDB) is the principal authority providing safe rural and urban drinking water. It falls within the MWSD, and has been implementing Asian Development Bank-supported projects.
- In NWSDB project districts, **Rural Water Supply and Sanitation Centres** (RWSSCs[1]) have been established in NWSDB offices, to support CBOs and *Pradeshiya Sabhas*, and also to provide advisory services to the general public.
- The **Community Water Supply and Sanitation Project** (CWSSP) supported by the government and the World Bank, and housed until recently in the **Ministry of Urban Development and Water Supply**, also implements rural water supply projects, and is the second major model of community-based rural water supply provision. It works through a central office, district-level **Rural Water Supply Support Units** (RWSSUs) and **Rural Water Supply Support Cells** (RWSSCs[2]) in *Pradeshiya Sabhas* offices. The CWSSP formally ended in December 2010.
- Responsibility for water service provision is vested with the ***Pradeshiya Sabhas***, the lowest level of democratic government, although CBOs oversee scheme construction and maintenance. In CWSSP districts,

Pradeshiya Sabhas have been strengthened with a small **Technical Cell** to provide post-construction support.
- **CBOs** operate and maintain village schemes on behalf of the community, designs tariffs and collect money, but do not own the assets – and hence cannot raise a bank loan using assets as collateral.

Service Delivery Models

The rural water sector has gone through three broad phases: 1) provision through local authorities (1948-1975); 2) technology-oriented provision through the NWSDB (1975-1993); and 3) community-based provision since 1993 through local bodies, supported either by the NWSDB or the CWSSP. Sri Lanka had two principal SDMs recognised in sector policy:
1. The NWSDB model: the NWSDB designs schemes, using government and external funding to construct the systems (either alone or through CBOs), and then hands them over to CBOs for O&M – although asset ownership remains with the NWSDB.
2. The CWSSP model: construction is outsourced to the private sector, but the CWSSP Project Management Unit has an intensive capacity building and awareness generation programme and a strong decentralised approach to implement a comprehensive package of sanitation, rainwater harvesting, hygiene awareness, environmental conservation and income generating activities.

In 2010, the two models were unified under a single Ministry, making the implementation approach more cohesive.

Key issues

With high coverage rates, Sri Lanka can be said to have 'scaled up', but it faces a new generation of problems, including source sustainability, institutional role clarity and rising demands for better quality water. Government agencies such as the NWSDB are facing a challenge to transition from being a provider to a facilitator, and to enable full decentralisation to *Pradeshiya Sabhas* and, through them, to CBOs. This is especially true when it comes to retrenching or seconding the large technical staff in the NWSDB to either *Pradeshiya Sabhas* or CBOs.

Thailand

Thailand is a lower middle income country with a GDP per capita of US$8,060 (IMF, 2010), making it the second largest economy in Southeast Asia. The country has seen remarkable progress in human development in the last 20 years, and it will probably achieve most, if not all, of its MDGs well in advance of 2015. Decentralisation in the last decade is one of the key factors that has shaped the administrative system, as well as public services including rural water supply. For the past four decades the government has given water supply high priority, with about 90% of the population now having access to safe water – a higher rate of whom are in urban areas.

Administrative set up

The state administrative structure is made up of three systems: **central administration** (ministries and their departments), **local administration** (in provinces and districts), and *Tambon* **Administrative Organisations** (TAOs) (local autonomy).

Main sector institutions

- Water resources are administered and managed by eight ministries with different priorities and programmes that sometimes overlap or are in conflict. At the central level the main sector institutions are the **Department of Public Works** in the **Ministry of Interior** (MOI), and the **Department of Health** in the **Ministry of Public Health** (MoPH). The MOI is responsible for communities with a population over 5,000, and the MoPH is responsible for communities of 1,000-5,000.
- Piped water in Thailand is currently provided by three agencies, depending on area and population served. Urban areas are served by the two main state enterprises: the **Metropolitan Waterworks Authority** (MWA), a state enterprise under the MOI for Bangkok and its provinces, and the **Provincial Waterworks Authorities** (PWAs) for all other cities and towns. The MWA and PWAs both oversee and regulate, as well as provide technical support to TAOs and Village Water Committees.
- Rural villages and communes with populations under 5,000 are overseen by the **TAOs**, a local administrative division (on average, one TAO covers 10 villages). The process of decentralisation devolved power and responsibility for development planning and management for public services to the local level, the TAO. One hundred and eighty functions were transferred to the TAOs, including responsibility for the physical assets of rural water supply. TAOs have revenue raising powers, a broad range of local government functions (including rural water supply), and are taking on roles in oversight and subcontracting services to private companies.

- After construction, the systems are transferred to **Village Water Committees** to continue running the service, with the aim of becoming financially self-sufficient. Village Water Committees are an independent body representing water users. Most comprise four to six people, including the chief, vice chief, accountant, and system operator/maintenance person. Many of the Committees work on a voluntary basis, with the exception of the system operator who receives a moderate salary. Village Water Committees and TAOs receive training and technical support from two main ministries: the **Department of Water Resources (DWR)** and the **Department of Local Administration**. However, owing to decentralisation and a change in the budgetary system, technical support from central government is declining.

Service Delivery Models

Thailand has two principal SDMs for rural areas recognised in sector policy:
1. Community self-supply: by rainwater harvesting and storage in family water jars.
2. Piped water supply systems: currently piped water is provided by the three main agencies: MWA, PWA and TAOs, with Village Water Committees.

Key issues

Most rural people use at least two water sources: rainwater from jars and tanks, and shallow ground water from tube wells. However, increasing numbers of villages are getting piped connections to PWA systems. This means that a significant number of people have access to three water sources. Self-supply is an important approach and is an accepted part of the solution for rural areas, with institutional support. It is mainly focused on water for drinking. CBM and support functions have built-in flexibility for management options from the more simple community management up to full private delegation. It is an institutionalised response to the one-size fits all mentality seen in many countries, and it helps to deal with localised capacity constraints at decentralised levels. However, it is unclear whether and to what extent rainwater harvesting, traditionally practiced in most rural communities, is formally recognised when piped water systems are planned and built in each village and commune.

Uganda

Uganda has a population of just over 30 million, over 80% of whom live in rural areas. Economic reforms since 1990 have resulted in strong economic growth based on Uganda's focus on investment in infrastructure, lower inflation and better domestic security. The global economic downturn has hurt Uganda's exports however, although Uganda's GDP growth is still relatively strong due to past reforms and sound management of the downturn. Uganda is classed as a less developed country with a GDP per capita of US$1,196 (IMF, 2010).

Administrative set up

The country is divided into over 100 districts, and each district is then divided into sub-districts, counties, parishes and villages.

Main sector institutions

- At the national level the **Ministry of Water and Environment** (MWE) has overall responsibility for initiating the national policies and for setting national standards and priorities for water development and management.
- The **Directorate of Water Development** (DWD) is placed within the MWE and is the lead agency responsible for managing water resources, coordinating and regulating all water activities, and providing support services to local government and other service providers.
- In the process of decentralisation, **Technical Support Units** (TSU)s, were set up to provide *ad hoc* strategic capacity building to **district local governments** (DLGs). The four major regions of the country are divided into eight sub-regions constituting eight TSUs, each headed by a Focal Point Officer.
- The **District Water Office** (DWO) is the lead office for the water sector at DLG level. Its main responsibilities are planning and implementation (e.g. procurement, contract management, reporting and accountability). The DWO is also responsible for capacity building, as well as ensuring O&M of water facilities through private operators for rural piped water schemes and Water User Committees (WUCs) for rural point water sources.
- The **District Water and Sanitation Coordination Committee** (DWSCC) provides a platform for coordinating and harmonising of approaches in the implementation of the activities of the rural water and sanitation sector in the DLG, and strengthens collaboration across sectors and between different players.
- In Rural Growth Centres (RGCs) and Small Towns, **water user associations** (umbrella organisations) and sub-county Water Supply and

Sewerage Boards (WSSBs), through contracted private operators, are the major actors for O&M.
- **WUCs** are responsible for day-to-day O&M and administration of point water supplies and, in some cases, gravity-fed systems.

Service Delivery Models

There is a policy framework defining and specifying Service Delivery Models for rural areas, RGCs and urban areas, each taking a different form. These are largely, though not exclusively, linked to technology options and settlement type. The two principal Service Delivery Models recognised in sector policy are:

1. WSSBs, through contracted private operators for Small Towns and RGCs: there are two alternatives – handpump mechanics and scheme attendants who provide maintenance services to water users in rural and peri-urban areas, operating as private entrepreneurs and system caretakers, respectively.
2. Community Based Management System (CBMS) for rural water supply: the CBMS model has a number of limitations, but currently is considered the appropriate option for what are mainly point sources (handpumps and springs) in rural communities. It is recognised that efforts should go into 'professionalising' the CBMS.

Self-supply initiatives are promoted and recognised, but are not formally stated in any policy statement. The self-supply approach is complementary to conventional CBMS, whereby government or NGOs pay for between 90 and 100% of the cost of the physical infrastructure. The MWE has now embarked on the process of developing a comprehensive framework for self-supply. It tries to set out the roles and responsibilities of various actors, and calls for further definition of the possible technologies required.

Key issues

Real effort is being made to decentralise rural water supply to DLG. A comprehensive programme has been established with clear responsibilities for service providers and DLG, through which the DLG receives support from central government. This is accompanied by a comprehensive financing mechanism. The advanced SWAp, in which government and key development partners pool funds for one WASH programme, has resulted in reduced fragmentation of effort, and better alignment with government policy and approaches.

United States of America

The USA is one of the wealthiest countries in the world, but it is also characterised by large and growing disparities: the Gini-coefficient is 45, up from 35 in the 1970s, and higher than all other industrialised nations. In total, not including the U.S. territories, roughly 1.8 million people lack access to improved water supply.

This number translates into 1.15 million urban and 600,500 rural residents. Those who are the least likely to have adequate services are the poor, elderly and people living in rural communities. In the presence of relatively professional providers and nearly universal coverage, the role of state and national government is limited to regulation (with a focus on private operators), control, and certain aspects of financing; local government is very often the system owner.

Administrative set up

The governance structure is federalist, and largely decentralised, with 50 individual states having autonomy on most matters. These are further sub-divided into 3,143 counties.

Main sector institutions

- The **Environmental Protection Agency** (EPA) provides the drinking water quality (public health) regulatory framework under the Safe Drinking Water Act of 1974. Under the federal system, institutional frameworks are largely set at state or sub-state level.
- There are **state regulatory agencies**, known as 'primacy agencies', responsible for ensuring water is delivered and which regulate and monitor compliance with health and environmental regulations and convene all financing agencies at the state level to determine the needs within the water sector. Along with the water system financing office of the **Rural Utilities Service (RUS)** of the **United States Department of Agriculture** (USDA), these primacy agencies have offices at county or parish level in most states. EPA, USDA and the state primacy agencies work closely with not-for-profit technical assistance provider organisations to ensure that rural residents have potable water services.
- The management of rural water supply is done through a multiplicity of organisational types that have the responsibility for planning, implementation and administration, including **municipal or state rural utilities, tribal or federal water utilities, private companies** and CBOs.
- Depending on the ownership type, the day-to-day operations of running a rural water system may be contracted out to **rural water operators** who are either employees of the community, the district, a private contracting company, or a private water entity.

Service Delivery Models

There are four principal SDMs:

1. The community model: serves a community (town, village, or hamlet) with multiple types including public municipal utilities, not-for-profit community models and regional water systems.
2. The PPP model: in rural areas the most common choice is not-for-profit utilities (CBOs which are recognised as service providers), with the option to have a private company as the operator.
3. Fully private options: these include CBM where the provider always owns the assets.
4. Household self-supply: this continues to be a major component of rural water supply, generally through wells constructed by home owners.

Key issues

High quality service provision is enabled by high levels of subsidy and loan mechanisms (for which service providers may access different and changing channels/sources). These enable capital maintenance and a strong network of professional support, even to small rural operators. Post-construction support, technical advice and capacity building are provided by two principal organisations – the Rural Community Assistance Partnership (RCAP) and the National Rural Water Association (NRWA), which organises the circuit rider program of technicians who provide post-construction support to rural operators. Both are 'bottom-up' organisations providing support to members, but are equally well linked into government (funding) systems both at federal and state level.

The priority lies with improving performance of operators (e.g. through capacity development and regulation) and water quality. At the same time the rural water sector faces major challenges with an expensive and aging physical asset portfolio, and the risk of slippage, as many systems come to the end of their life span; current investment levels are insufficient to meet these replacement needs. Financing mechanisms come from a combination of local, state, regional, and national funding sources, and many rural populations rely on political leverage to access new funds for rehabilitation and replacement.

CHAPTER 4
Findings from the country studies

This chapter presents an analysis of the findings from the research studies in the 13 countries. In broad terms the findings from the studies are presented according to the same elements of the framework used by the country teams to assess the status of rural water services in each individual country. Therefore, the analysis is presented as follows:

- The first section 'The status of access to sustainable rural water supply services' looks across the countries to provide insight into how service delivery levels are defined and measured in terms of indicators and targets for sustainability.
- The second section 'Institutional arrangements for rural water supply' presents a review of rural water sector reform and the separation of roles and functions at different institutional levels, including the impacts of broader decentralisation processes.
- The section on 'Management options' provides an assessment of trends and factors affecting service delivery at the local level, including the way in which different management options have evolved.
- The section on 'Service authority functions' gives a review of the intermediate level in supporting service provision options by looking into key authority functions such as planning, monitoring and post-construction support.
- The last section, on 'An enabling environment for service authorities' assesses the national level, with a review of factors and elements making up the enabling environment in which service delivery takes place, and deals with issues of financing, accountability, regulation, learning, sector capacity development, and harmonisation and coordination.

The status of access to sustainable rural water supply services

Service levels

One of the basic building blocks of service provision is in having clearly defined levels of service within sector policy and norms; put simply, a definition of the type of service a consumer has (or should have) access to in terms of characteristics of quantity, quality, access, reliability and continuity. In some cases, definition of service levels is conflated with technology type, and while technologies have a bearing on the level provided, the two are not necessarily inter-changeable (as in the difference in service provided by a handpump located inside a single family compound, and that provided by a communal handpump serving hundreds of people).

All countries in the study have defined *minimum* service levels in their policy and legal frameworks; in other words, basic standards of service that need to be complied with, including the quantity of water to be supplied, often the distance between household and water point (in case there are no household connections), and water quality standards to be complied with. Many of these are in line with the World Health Organisation definition of access to water supply, which is defined as access to an improved source of water of at least 20 lpcd within one kilometre of one's dwelling (WHO, 2010b). A country's own definitions may deviate from this in some aspects of service, e.g. in the minimum amount provided, or the maximum distance between household and water source. Some may have their own water quality guidelines, but the basic definition of service level is structured along essentially the same lines. When assessing these definitions, a number of remarks can be made on how service levels are defined and understood in the sector:

- In some cases, service levels are further defined according to technology option. For example, **Burkina Faso's** norms state that distance to a public tap in a small piped system should be less than 500m; but for handpumps it can be less than 1,000m. Likewise, handpumps are expected to provide at least 20 lpcd, while piped systems should provide between 40-60 lpcd. In some respects this approach to setting out service levels could be seen to reflect the water ladder approach of WHO/UNICEF (2010).
- Some countries have defined other aspects of the service level such as reliability, continuity of supply, or 'crowding' – which describes the number of people expected to share a handpump or public standpost.
- Some countries have defined minimum service levels, but hardly ever apply them in practice. For example, **Honduras** follows the JMP definition, but piped water supply systems with household connections are the norm in water services provision. Curiously, the specific service levels for such systems are not clearly defined in Honduras. At most, there are design norms for piped systems in different sizes of rural settlements; a similar approach is followed in **Colombia**.

Table 4: Access to rural water supply services

Country	Use of improved water supply total	Use of improved water supply in rural areas	On track for MDG rural water supply
Benin	75	69	On track
Burkina Faso	76	72	On track
Colombia	92	73	On track
Ethiopia	38	26	Not on track
Ghana	82	74	On track
Honduras	86	77	On track
India	88	84	On track
Mozambique	47	29	Not on track
South Africa	91	78	On track
Sri Lanka	90	88	On track
Thailand	98	98	On track
Uganda	67	64	On track
USA	99	94	On track

Source: WHO/UNICEF, 2010

The most commonly used indicator for (rural) water supply is access to (and use of) improved water supply sources, and routinely compiled by the JMP[1] (see Table 4 for the indicators for the study countries).

The countries in the study reflect some general trends observed in the most recent JMP assessments' report. Access to improved water supply in rural areas continuously lags behind urban areas. Most countries are on track for achieving the MDGs for water supply; those which are off track are nearly all in Africa.

Another partially related trend observed in various countries is an increased demand for higher levels of service, either in the form of piped water supply or household options through self-supply. This was also noted in the JMP Report (WHO/UNICEF, 2010), which reports that more than 1.2 billion people worldwide gained access to a piped connection on premises, this being more than twice the population that gained access to other improved drinking-water sources. However, this trend is mainly confined to middle income regions such as East and Southeast Asia, Latin America and the Middle East. In Africa and South Asia other levels of service remain more important. The underlying factors for this demand are manifold, but include populations getting richer and having higher aspirations, in some cases stimulated by (re)migration and remittances from urban areas to rural areas. Returning migrants from urban to rural areas in countries like **Thailand** and **Colombia** have been reported to contribute to an increase in demand for better *quality* water, and for household connections in **India** and **Burkina Faso**. The desire to have similar facilities in rural villages lies behind this demand. Remittances have been reported

Table 5: (Proxy) indicators in use for sustainability of rural water supply

Country	Proxy formal indicator for sustainability of rural water supply	Value	Source
Benin	Functionality of water facilities	73% (handpumps) 79% (springs) 52% (dug wells) 69% (small piped systems)	Adjinacou, 2011
Burkina Faso	Functionality of water facilities	82% (handpumps) 66% (small piped systems)	Zoungrana, 2011
Ethiopia	Functionality of water facilities	67%	Chaka et al., 2011
Ghana	Functionality of water facilities		Case studies report functionality of boreholes varying from 58 to 90% (Skinner, 2009; and Bakalian and Wakeman, 2009 respectively)
Honduras	Composite indicator classifying performance of service into four levels	78% not classified as requiring major intervention	SANAA, 2009
India	Extent of slippage[A]	30%	GoI, 2008
Mozambique	Functionality – for handpumps only	85%	DAR/DNA, 2010
Uganda	Functionality of water facilities	81%	MWE/DWD, 2010

[A] While the concept of slippage is not an indicator but rather a descriptive term used by Indian authorities, it is a useful guide to the state of rural water services.

to be used for upgrading of basic water supplies, or for self-supply, in the form of household boreholes, even though there is little quantitative information on the relative size of these investments.

Sustainability definitions, indicators and targets

Unlike for coverage or access to water, there is no globally agreed definition for sustainability of rural water supply. In the study countries we have found a number of ways of expressing (proxy) indicators for sustainability of rural water supply (see Table 5).

On the basis of in-country knowledge of existing rural water services some of the values quoted by official sources are questionable; for example, the relatively high figure for functionality of handpumps in rural **Mozambique** quoted by the Rural Directorate. However, what the Table does indicate is that different types of indicators are being used to measure sustainability, reflecting different ways of conceptualising this dimension of a service. These include the following:

Functionality. The proxy indicator most commonly found, particularly in sub-Saharan Africa, is functionality, which expresses the percentage of water points working at any given time. Functionality is normally measured by a one-time check on a water facility or water point to determine whether the system is working at the time, and is normally a binary condition (yes/no). This indicator can work with simpler point sources (handpumps) where the system either works or it does not. But it may not be as useful when applied to more complex piped water systems which generally do not fail completely, but rather show gradual decreases in performance.

There is another more fundamental limitation to this indicator as it only looks at the 'output', and not at the underlying factors that may make a service sustainable or not. So, for example, functionality on a given day may indeed be 'zero' or sub-optimal, but in a case where many of the elements are in place (i.e. a strong operator, good tariff recovery, well coordinated support and monitoring, etc.) it is likely that the service will indeed be sustainable over time. Or, conversely, a system may be functional on a given day, but at the same time present many risks when these other elements are not in place. So, there are risks in using functionality too narrowly as an indicator of sustainability. Nevertheless, because functionality is relatively easy to measure, when assessed at regular intervals it remains a useful indicator, and when taken periodically over time, 'snapshots' of functionality can give a measure of sustainability.

Slippage. India uses the concept of slippage, referring to the slipping back of a village from a fully covered status to a partially or non-covered status (GoI, 2008). Slippage could be due to problems of source functionality, water quality and population growth. Rather than focusing on the service provider, it focuses on the access to the service by consumers. It reflects a deeper understanding of service delivery than the simple on/off indicator. However, it also shares some of the limitations of the functionality indicators in that it does not allow understanding of the underlying factors affecting sustainability. For example, a handpump may have only a small fault which a mechanic is due to fix. If the fault is dealt with quickly then slippage will not be much of a problem.

Composite indicators. In countries where more complex piped systems are the norm, such as **Colombia** and **Honduras**, composite indicators are used; a similar system has been developed in Bolivia by the Association of Municipalities of Cochabamba. These normally assess not only the status of the service provided (whether it is functional or operating at a sub-optimal level), but also key characteristics of the service provider such as, for example, the status of its financial records, or the relation between water committee and consumers. The scores on these sub-indicators may then be grouped into overall categories, such

Box 1: Categorisation of sustainability of rural water supply systems in Honduras

All rural water supply systems in Honduras are visited with some frequency by technicians. These update a fact sheet with key indicators of the systems, including water quality, technical state of the infrastructure, presence and activities of the water committee, and financial management. These are then grouped, through a software called SIAR into four categories: A, B, C and D. Depending on the category, recommendations are given to address the points on which it scores low. The formal definition of the four categories is as follows:

Category	Status of the system	Recommended intervention
A	The system functions well and there is potable water every day. Water is treated with chlorine. There is a water committee which meets regularly, and an operator carrying out O&M tasks.	Activities geared towards optimising community participation and continued strengthening of management tasks by the water committee.
B	The system may be working but there are management gaps that may put the sustainability at risk. No investment in infrastructure is required to move to category A, but should be geared towards strengthening the capacity of the water committee.	Supporting and strengthening management capacity. Supporting accountability and participation of the users.
C	The system may function only partially but there are management and physical deficiencies that put the sustainability at risk. Infrastructure investment is needed to move to category A, but that can be done with existing funds of the community.	Same as B, but support to the water committee in defining the work that need to be done, their budgeting and identifying of sources of funding.
D	The system is in such bad management and physical state that the costs of improving it and bringing it to category A, are beyond the possibilities of the community. Its life span may be over.	Define feasibility to be considered in future investment plans.

Source: SANAA, 2009

as the four categories identified in **Honduras** (see Box 1). This system allows identifying and anticipating risks to sustainability, even when the physical system is not manifesting any problems yet. For example, if a CBO is not collecting tariffs adequately, it may be predicted that system performance will deteriorate as basic maintenance functions cannot be financed.

Such indicator systems are gaining ground. Godfrey, et al. (2009) report on the use of a simplified sustainability check, using a scoring system for five underlying sustainability factors, in a number of districts in **Mozambique**. This also allows for the identification of the underlying factors that are most problematic in a given area, or even for a specific water point, which in turn allows for corrective action. The system in **Honduras** even includes a set of generic follow-up actions as part of its sustainability indicators. One major disadvantage of this kind of indicator is that it requires more resources for data collection and analysis. To be useful for planning purposes this implies not only an information system, but capacity at local and higher levels to take short-term and longer-term management decisions and follow-up actions based on the data collected.

No indicator. Curiously, in five of the country studies there is no information available on the status of sustainability of water supply services, as there is no indicator for this, nor collection of information on this topic.

In addition to these four categories of indicators, the literature also suggests other indicators for sustainability. These all have a stronger time element. One is for example, the measurement of the functionality of systems 10 years after project completion. This would give a good indication of the resulting sustainability of a service. An interesting example is the concept of water-person-years, which indicates the number of years a service is working times the number of persons supplied (Koestler, et al., 2010). However, these are indicators that allow looking back at how sustainability has worked after a time, rather than actual tracking of current performance.

Despite differences in definition of indicators, and probably in the reliability of some of the indicators, the results show an average of 20-40% of water services are not functional nor delivering a sustainable service. These figures correspond to other data sources. For example, a study by United Nations Children's Fund (UNICEF) reported on in RWSN (2009) showed an average level of non-functioning handpumps of approximately 36% in 20 countries in Africa (see Figure 6). Other studies and sources reveal similar levels. Figures indicate an 'all **India**' slippage rate of more than 30%, but with some states showing as many as 60 to 70% of rural facilities as having fallen back to partial coverage or no coverage at all (IRC, 2009). Probably, if composite indicators would be applied, levels of non- or under-performance would be even higher. In **Honduras** a reasonable 35% of the systems were performing optimally (classified as A), 43% presented problems (either in category B or C), and the remainder required major intervention (in D). Recent work from **Mozambique** revealed that most water points scored between 50 and 75% in their sustainability test, indicating medium to high risks of non-sustainability (Godfrey, et al., 2009).

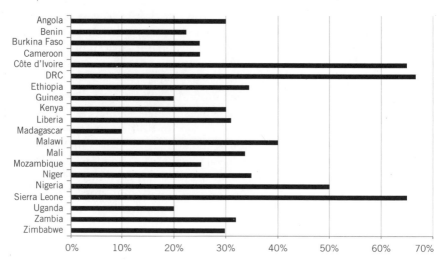

Source: RWSN, 2009

Figure 6: Estimated % of broken handpumps

These figures are at odds with the official sector claim of 85% functionality for handpump based systems (DAR/DNA, 2010).

Targets for sustainability

Despite the fact that various indicators exist, and that eight of the countries in the study collect and use indicators for sustainability, only two of these have explicit sector targets for sustainability: **Honduras** and **Uganda**. Both of these countries have set quantitative targets to which to elevate the different categories and functionality levels respectively. When **Honduras** started its sustainability programme in 1986, a rapid survey showed that only 7% of the water systems could be classified as A (Rivera Garay and Godoy Ayestas, 2004). On an annual basis targets were set to increase this by a certain percentage. 2007 was the last year that a target was set to increase the percentage of systems classified as A from 38 to 41% – a target that was met. With the significant reduction of the programme, targets have been abolished, and the latest count showed 35% of systems classified as A. In **Uganda** the target for functionality set by Government of Uganda for the financial year 2009/10 was 86%, and that for 2014/2015 is 90%. According to the Sector Performance Report (MWE/DWD, 2010), functionality dropped from 83% (in 2008/09) to 81% (in 2009/10). **Mozambique** plans to include a similar functionality target in the performance indicators for the System Wide Approach (SWAp) that it is setting up for rural water.

Discussion: indicators for sustainable service delivery

The findings presented in this section show a mixed picture of the importance given to sustainability as a formal sector indicator or target. Current approaches to monitoring in many of these countries focus primarily on the number of systems built and the numbers of people served by these systems. This echoes the importance given to tracking progress in increasing coverage through, for example, the JMP.

Less structured attention is given to whether or not these systems actually continue to deliver reliable services. Obviously, the lack of a clearly defined indicator or target for sustainability does not mean that a sector is not paying attention to it – in spite of the absence of a sustainability indicator, the USA is putting efforts into structurally trying to sustain rural service delivery. But formal targets or clearly defined proxy indicators allow a sector to take more concerted efforts to improve sustainability, measure progress towards it and take corrective action in a targeted manner. Clearly elaborated sustainability indicators, with corresponding sector targets, would therefore be a first key constituent element of establishing an SDA. Composite sustainability indicators, that contain both information on the service provided and the service provider, are best able to reflect the complexity of sustainability and anticipate sustainability problems.

However, collection of data for measuring sustainability in and of itself only brings us so far and the real benefits will only accrue when there is the corresponding capacity to analyse such information and act upon it, through improved performance management. Put simply, the effort of collecting data through carefully selected indicators is only really useful if remedial action can be taken to decide how and where to invest resources, to provide back-stopping support, or to look into particular issues or trends in much more detail.

Institutional arrangements for rural water supply: sector reforms separating roles and functions and decentralisation

Almost all developing countries are, or have been, going through broad ranging processes of decentralisation of governance functions over the past 10 to 20 years. This phenomenon transcends many different sectors (health, education, as well as WASH), and has often been encouraged and supported by donors and driven by the notion that decentralisation will bring greater levels of involvement of, and accountability to, the recipients of services. This shift has made the decentralised governance unit the critical building block for rural water service delivery in almost all countries in the study. Decentralisation is a political process which is external to the rural water sector, but has a profound impact on the institutional roles and responsibilities and other accountability functions. In almost all cases decentralisation has gone together with structural reform of the water sector, often involving the separation of functions previously held by one monopolistic state entity (i.e. policy, planning, implementation and oversight). Therefore, understanding of sector reform is closely linked to an understanding of broader decentralisation processes and trends.

Decentralisation processes

The key process through which the reforms have happened is decentralisation, meaning the transfer of authority functions and financial resource allocation to the local level and, in increasing numbers of cases, the ownership of water system assets. Decentralisation is based on the subsidiarity principle, whereby decisions regarding the use and management of water are made through the involvement of local level stakeholders. This process has not automatically meant that local authorities take over the responsibility for the actual provision of services in rural areas, rather they are typically given the responsibility of **guaranteeing** service provision, through tasks such as planning, financing, implementation, monitoring and support of rural water supply. The extent to which decentralisation has happened, and the support provided to this process of transition of authority, has resulted in very different experiences across the country cases included in this study. A number of trends emerge which illustrate that decentralisation happens at different speeds and with different levels of support to the newly mandated service authorities. The following types of experiences can be identified:

1. **Phased decentralisation:** A number of countries have followed a two-step approach with first deconcentration of certain functions towards the provincial level, followed by a further effort to decentralise to local government. In **Benin**, where the decentralisation process started in the 1990s, a strong presence of deconcentrated state entities was established at departmental level for many years, before a more recent decentralisation to the **communes**. In **Mozambique** the decentralisation process has been extremely slow to develop, starting through legislation from 1997 onwards; the actual transfer of capacity, financing and real decision-making power is still very limited.
2. **Partial decentralisation:** This refers to cases where different degrees and dimensions of decentralisation are employed in parallel. In **Ghana**, for example, despite the existence of the legislative framework and the establishment of the DAs, there has been a generalised lack of transfer of staff and resources to lower levels. The result has been that staff who should fall under local government still report vertically to the line ministries, causing confusion about roles and responsibilities. This has been compounded by the 'projectised' way in which the reform process has been implemented, in which different donors take on some elements of the reform process in different projects, supporting some regions and districts, but not others. In some respects this situation shares aspects of the reform process in **Honduras** where planning and implementation functions have been decentralised to municipal government, but where support and capacity building efforts have been poorly coordinated and

funded. The vacuum of weak coordination from central level, and the lack of a common sector investment plan or strategy, has been filled by 'projectised' support to some municipalities, but not to all.

In the larger countries with federal contexts, such as **India** and the **USA**, although decentralisation is in place as a national policy, there are wide variations in the speed and extent of progress, with individual states having the autonomy to pursue differing approaches under a federal system. For example, in **India**, there are some states with hardly any progress on meaningful decentralisation (such as Orissa), and in others (including most notably Gujarat and Maharashtra) there are innovative experiences which have been rolled out at scale. In the case of **Ethiopia** strong political power has allowed for decentralisation process 'on paper' driving responsibility for service provision to the *wordea* and *sub-wordea* (or *kebele*) levels, but in practice these entities rarely function as real service authorities due to chronic under-funding.

3. **Inadequately resourced decentralisation:** This refers to those countries where decentralisation has happened quite rapidly, at least on paper, but without adequate support or where certain capacities for local authorities have not been decentralised. This has occurred, for example, in **Burkina Faso** which accelerated the process in 2006 by creating the possibility of **commune** authorities with the mandate for guaranteeing delivery of rural water services. This has included fiscal and administrative decentralisation, but with limited resources. As a result, many **communes** are woefully under-resourced, staff capacity is very low, and the systems of local government have not been able to keep pace with the reform process; for example, **communes** are still not yet legally able to open bank accounts. Whilst seeming to appear extensive – say in comparison with the 'messy' processes in **Ghana** and **India** – the reality in **Burkina Faso** is that little may have changed at the decentralised level in the majority of cases.

4. **Wholesale planned decentralisation:** In a number of the countries in the study there has been relatively well-planned and complete processes of decentralisation that have helped to progress the sector. **Colombia** had a long and quite successful process of decentralisation, alongside the establishment of community management as the primary model for rural service provision. **South Africa** has witnessed one of the most systematic processes of sector reform with key targets and process indicators to accompany broader decentralisation. As a result, there was a quite quick transfer of resources, budgets and decision-making power, but also with relatively good support and follow-up. In reality there continues to be severe challenges to decentralisation, even in such a relatively rich country. **Uganda** has one of the most advanced and complete processes of decentralisation and reform in place among the country studies. Decentralisation efforts began in the mid 1980s as the country emerged from a prolonged period of internal armed conflict, and initially were not 'donor-driven' as has been the case in so many

countries on the continent. Administrative and financial dimensions of decentralisation are relatively well established in **Uganda**, but true political (decision-making) autonomy remains limited.

Reform to functions in rural water service delivery

A key part of the reforms that have accompanied broader decentralisation has been the separation of roles and functions related to service provision for rural water. When VLOM and community-management approaches formed the main paradigm in the 1990s, many rural water supply projects, often funded and implemented by donor programmes and INGOs, would work directly with communities, many times by-passing (local) government or, at most, keeping them informed only. But with a more articulated role for local government in roles such as planning, financing and monitoring, this had to change and, in response, rural water supply programmes started to give a more prominent role to local government from the early 2000s onwards. More recent steps in sector reform have included support to small-scale private operators, particularly for more complex piped systems in larger rural settlements and small towns. Promotion of PPP has been particularly strong in parts of West Africa (**Senegal**, **Mali**, **Niger** and, to some extent, **Burkina Faso**) as well as in **Uganda** which has supported programmes in a number of RGCs.

Above all, efforts have been taken to separate the service authority functions, such as planning, implementation, monitoring, oversight and post-construction support, from the direct service provision role. The latter functions have been transferred to different types of service providers, above all in the form of community management organisations, but also private operators and households (see section 'Management options as part of Service Delivery Models', p73). The former role – of service authority – has been decentralised almost exclusively to local authorities, at least in theory; but in practice the reality is more one of a mixed picture. In the wealthier, less aid-dependent countries such as the **USA**, **Thailand**, **India** and to some extent **Colombia** these functions genuinely have been transferred to lower tiers of government. In **Uganda** following the establishment of the Decentralisation Act in 1995, the water and sanitation sector benefited from a long and relatively well-structured process of devolving mandate and powers to district level. **South Africa** stands out as one of the most complete examples of separation of functions between the municipal WSA and the range of entities that are responsible both for capital investment projects and day-to-day management of services (the WSPs).

But in many other countries – particularly where interventions still tend to be donor or project driven – the reality is that many functions have not been decentralised, and are retained by other centralised or deconcentrated actors (including large-scale programmes), and local government has limited capacity to act independently. For example, in **Benin**, the departments still play an important function in channelling funds and procurement, thereby supplanting certain functions intended to be carried out by the communes.

FINDINGS FROM THE COUNTRY STUDIES   69

In **Ethiopia** it is still the donor agencies and INGO-financed programmes that provide the bulk of both capital investment activities and support services. Local government capacity remains chronically low, with greater government capacity found in the regions where the deconcentrated offices of the central Ministry are located (the Bureaux of Water Resources Development [BoWRD]). Another example is **Mozambique** where some authority has been deconcentrated to provincial government, but this is still limited, for example, provinces have more of a role in planning and monitoring through the DPOPH. Certain functions are being decentralised to district authorities, but these are still not fulfilling the function of a service authority, but are seen to be more of an implementing partner and communication channel between communities and provincial government. In **Ghana** donors also continue to channel funding through the centralised CWSA rather than through DAs, although this is starting to change.

Unlike in the urban water sector, formal regulation of rural services is still very limited, with only a small number of countries having a regulatory authority at all for the sector. For rural and small town services this function is rolled into the more general activities carried out – again in theory – by local government as part of the service authority functions (see section 'Accountability, regulation, oversight and enforcement', p94). Alongside this separation of levels of decision-making and control, there has been an increasing separation and specialisation of functions according to the different stages in the life-cycle, for example, having dedicated agencies for project implementation, service provision and, in some cases like **Honduras**, post-construction support. In practice, the long-term support functions are often left to local government, but without sufficient resources to fulfil these obligations. Such reforms have had the aim of bringing in more checks and balances, accountability and specialisation and, in many cases, that has been achieved. But an inevitable result has also been a higher degree of fragmentation. This in itself need not be a problem if and when appropriate coordination exists.

One of the challenges to decentralisation, and concentrating service authority functions at district level, has been reaching economies of scale. Certainly, working at the level of the district or municipality provides an improvement over community-level interventions, as is often the case with NGO or even large-scale donor funded programmes. However, in practice, many local governments remain weak and poorly equipped, and there is evidence that some functions are being re-centralised to higher administrative levels. This has happened formally in **Colombia**, with planning and financial allocation taking place at the departmental level, and informally through the creation of groupings of municipalities (*manocomunidades*) in many other Latin American countries which seek to spread the costs and economies of scale for professional capacity across a number of municipalities. This has also been an area of tension in the dynamic between the push for decentralisation and the resistance to change illustrated by some agencies who stand to lose in the new decentralised order – for example, according to the CWSA in **Ghana**, the district level does not allow for economies of scale, and still continues to tender for multi-district contracts.

Finally, the transfer of assets has sometimes been a difficult issue as part of sector reform processes. While in some cases ownership of infrastructure assets is clear (such as **Ghana** where this is the property of the DAs, and authority is temporarily vested in local WSDBs to manage services), in others, such as **Mozambique**, final title is interpreted differently by competing ministries or departments (the Ministry for Local Government disputes ownership claims by the National Directorate for Water). In **South Africa** a major point of contention has been the transfer of assets from DWA to municipalities. Many of the assets were in poor shape, and municipalities were not always prepared to take them on. During the transfer of assets there was wide consultation between the Department of Provincial and Local Government, the South African Local Government Association and DWA to resolve some of these issues. Likewise in **Burkina Faso** assets were only legally transferred from the central Department of Water under the Ministry of Agriculture to communes in January 2010. In theory the ministry should have rehabilitated all systems, many of which are in poor repair, before transferring ownership. But instead they transferred them in their current state, with only the equivalent of some US$2,250 per commune as a fund to rehabilitate systems (a totally inadequate sum).

Reforming agencies

The reforms of recent years have not only affected local authorities. The role of centralised and/or deconcentrated agencies previously engaged with direct implementation and support in the study countries have also had to re-assess their roles. These include entities such as the state agency SANAA in **Honduras**, the various Public Health Engineering Departments (PHEDs) in **Indian** states, the CWSA in **Ghana**, the Department of Water in **Burkina Faso** and the 'old' Department of Water Affairs and Forestry (DWAF) in **South Africa**. With sector reform, and particularly with fiscal decentralisation, funding for rural water services – in theory – is increasingly being passed straight to local government level. The typical roles these centralised or deconcentrated agencies are now being asked to play are around facilitation, post-construction support, and technical advice and oversight, rather than large-scale implementation (i.e. system construction). However, the picture is mixed and, in some cases, they still have a role in implementation. For example, in **Sri Lanka**, the centralised state agency NWSDB is still directly involved in construction and/or managing contracting for new systems and major rehabilitations, as is the CWSA in **Ghana**.

However, the change of roles from what Wester (2008) calls the 'hydraulic mission' of massive infrastructure development to facilitation has encountered reactions of resistance and co-option. Most commonly, the studies highlight a 'resistance to change' on the part of individuals, often linked to the loss of control of resource allocation through centralised procurement and contracting. For example, in **Sri Lanka**, as part of the decentralisation process, some technical staff of the NWSDB have been placed at district level to play a new, more advisory role to local government and CBOs. This change, which now puts these local level organisations on the same footing as central government engineering staff, has met with strong resistance, and it is not clear how well this new role of facilitation will be taken up. These findings are not unlike what has been found in reform of large monopolistic public sector institutions (or 'hydrocracies', as Wester [2008] calls them) for irrigation or water resources management.

In **India** the decentralisation and sector reform process has placed specific attention on the role and mandate of the existing engineering-focused implementing agencies (such as the PHEDs) to adapt to new roles. Various strategies have emerged to manage this change process. For example, the creation of a new parallel organisation such as WASMO in Gujarat; alternatively the internal transformation of the existing PHED in Maharashtra by imparting new skills and capacities; and in Tamil Nadu through a pilot project using external consultants and focussing on the culture of the organisation and staffing profiles to address 'attitudinal' problems (see Box 2). In other states (e.g. Orissa) there continues to be conventional set-ups mainly led by PHED, with more traditional or infrastructure perspectives. This situation is reinforced by the role of the Executive Engineer (usually from the PHED), who leads District Water and Sanitation Committees, and who often resists or even derails efforts to move towards a greater SDA, and to promote the role of the community in decision-making. These different examples reflect the size and plurality of

---

Box 2: Change management works!

The Tamil Nadu Rural Water Supply Project of the Tamil Nadu Water Supply and Drainage Board (TWAD) carried out an interesting exercise which served to change the attitude of engineers, who themselves then constituted a change management group, to encourage more responsible and motivated functioning. A key motivational question asked in intensive workshop sessions was: "What will you tell your grandchildren when they ask: 'you were in charge of the water supply in the state – and this is the status?"

The intensive and frank discussions that followed resulted in an acceptance by the rural water supply engineers that change was needed. They came out with the Maraimalai Nagar Declaration which was a revolutionary vision for conventional engineers coming from a hierarchical, top-down and bureaucratic organisation, and served as the basis for the Tamil Nadu Rural Water Supply Project Pilot Project, where each engineer 'adopted' a village to implement his vision. A total of 145 villages were selected. An independent assessment in 2006 showed that there was remarkable change in these villages once engineers decided to work with the community. The community, in turn, responded to improved water service delivery (including budget savings of 40% on electricity charges and greater user satisfaction) by volunteering to contribute for the previously free public stand posts. In 30% of the villages, contributions of Rs.10 (US$ 0.20) per household per month were received for public stand posts.

Source: James, 2011a

experiences in sector reform in **India**, which include the entire range from unreformed to reformed and decentralised structures at state level.

Discussion: impact of decentralisation and reforms on conditions for service delivery

This mix of experiences from the case study countries highlights a number of important aspects that have a bearing on rural water service provision. Firstly, it shows that fiscal decentralisation is key for decentralisation to become more than a token process. In some cases significant funding flows do reach down to the level of district or local government. For example, **South Africa** and **Uganda** have established mechanisms to channel central budgetary allocations to lower levels of government (the municipal infrastructure grant and the conditional grant respectively), but even in these cases there are still serious short-falls in necessary financing for institutions and activities that are required to make the shift to an SDA. In most of the remaining countries, transferring responsibility for service provision, with inadequate financial allocations to the local level, will not allow local governments to fulfil their roles, for example, in countries such as **Burkina Faso**, **Mozambique** and **Ethiopia**. Similar findings from a recent WaterAid study across 12 countries indicate that over two-thirds of expenditure at local level is still outside of the local government's budget and direct control (WaterAid, 2008).

Secondly, support to decentralisation is essential. Even when responsibilities are reasonably clear and financial resources are provided, decentralisation will require support to local government for a long time – perhaps indefinitely. In some cases this has been addressed – deliberately or otherwise – through a two-staged approach which recognises the time needed to develop capacity in local government, and leaves a structure at provincial level to support local government. However, there is a risk of the 'chicken and egg' dilemma of cause and effect, whereby central governments resist the real transfer of resources and decision-making power to local levels because it is claimed that there is insufficient capacity in the first place; **Ghana** provides a good case in point.

Equally, there are different experiences with sector reform with some countries, generally wealthier and less aid dependent, managing to achieve a more complete separation of functions between service authority (generally held by local government) and day-to-day operation and management of services (either community managed entities or small private operators). However, in many cases the hoped for reforms – and clear separation of functions – exist largely 'on paper', with the reality that much has not changed either because of lack of capacity at the local level, or a certain degree of inertia, and even resistance, on the part of strong technical parastatal organisations.

Taken collectively, the evidence from the case studies suggests that there has been a general trend towards the 're-engineering' of the rural sector towards new frameworks which can better support the delivery of services.. However, in reality, many countries are still struggling with incomplete reforms, or reforms that are stalled, meaning that the desired step change to service delivery has not yet come about. The reasons for this slow progress are broader than the water sector and include inadequate fiscal decentralisation, lack of financing, capacity constraints, and stalled public administration reforms.

This should, however, not necessarily lead to the conclusion that such reforms are less successful than what went before (i.e. more centralised approaches with single agencies carrying out multiple functions). Rather, it is too early to say whether such reforms will yield markedly improved results and, as with the broader processes of decentralisation, that these are long-term processes with no guarantees of success. The desired results of decentralisation – greater accountability in governance, better local participation and improved efficiencies – are not likely to become evident in the short-term, and there is evidence to show that when the process is not fully implemented, or poorly managed, it is likely that things will get worse before they get better. Despite this, the picture is not all negative and, even where reform and decentralisation only partially work, things can be improved significantly with centralised service provision. A number of countries, including **Uganda**, **South Africa** and **Colombia**, have all shown promise in terms of sector reform for rural water services. In short, it is still too early to say whether these fundamental changes will bring about improved services; therefore it is important to look beyond the merits and difficulties of decentralisation per se, and assess how key water functions are fulfilled and can be further improved.

Management options as part of Service Delivery Models

Over the past 20 years or more, a range of different management options have evolved in rural water provision. Some of these are formally endorsed by government policy; whereas others have been supported by donor or NGO programmes. One important distinction to note from the outset is the definition and description of such arrangements over the life-cycle of a service. Although not always the case, this often means that the arrangements put in place to build the new infrastructure (or to carry out a major upgrade or capital replacement project) may not be the same as for the management of the day-to-day O&M of the system.

A number of formally recognised management options were found across the study countries (see Table 6), with a clear predominance of the **CBM** approach. In all cases it exists as a formal model, defined within government policy, even though community management entities do not always have a clear legal status in all countries. Other options have also been recognised, including **public sector management** (through municipal utilities or local government providers) and the growing involvement of small **private operator** arrangements (mostly in the form of delegated contracting through PPPs). Finally, there is **self-supply** which is understood as the investment in and management of household facilities by the same households. There are many

Table 6: Formally recognised Service Delivery Models

Country	Community-based management	Delegation to private operators (PPP and NGO Operators)	Local government as provider	Self-supply	Urban utility (public, private or mixed)	Observations
Benin	✓	✓				Range of delegation options available depending on technology type and settlement size
Burkina Faso	✓	✓			✓	ONEA holds lease contract in seven municipalities. One case of variation to CBM, with an Association of Water Schemes covering 41 individual water user associations
Colombia	✓	✓	✓	✓		Formally recognised and regulated, distinction between four different CBM options which are well defined
Ethiopia	✓		✓	✓		CBM includes WSSBs for multi-village schemes. Self-supply only recognised in formal policy from 2009
Ghana	✓	✓		✓	✓	CBM is predominant model in small rural communities with a growing trend towards PPPs – DAs are always asset holders in the case of CBM
Honduras	✓		✓	✓	✓	Recognition that larger communities should get more professionalised CBM
India	✓			✓		Asset ownership is unclear – although *Gram Panchayat* is recognised as the lowest level of government, ownership does not have a legal backing at this level
Mozambique	✓					Formal status of water committees unclear but unlikely to be recognised as legal entities
South Africa	✓	✓	✓			Well-established process for determining type of service provider, but heavily skewed against CBM operators
Sri Lanka	✓	✓				Two different models for CBM, differing mainly in the capital intensive phase

Continued ▶

Continued

Service Delivery Model Country	Community-based management	Delegation to private operators (PPP and NGO Operators)	Local government as provider	Self-supply	Urban utility (public, private or mixed)	Observations
Thailand	✓			✓		Self-supply is an important approach, mainly based on rainwater harvesting (for drinking) with institutionalised support
Uganda	✓	✓		✓		Relatively long experience in testing small-scale PPP arrangements in RGCs
USA	✓	✓	✓	✓	✓	In rural areas most common option is not-for-profit utility or CBO, with the potential to have a private body as operator; in many cases provider also owns assets

other forms of management arrangements which are not formally sanctioned in government policy, which of course does not mean that they are not important; for example, self-supply is *de facto* an important service provision option in all countries, regardless of whether it is recognised and supported officially.

Community-based management

The CBM approach has its roots in the international Decade for Drinking Water and Sanitation (the 1980s), which ushered in a new wave of donor and NGO programmes, largely by-passing government structures in favour of communities and grassroots organisations. Part of the philosophy behind this trend was to place responsibility for O&M onto the community. One of the early manifestations of this approach was the development of VLOM technologies[2]. Although the decade yielded success in a massive expansion in coverage, it failed in achieving its target of 100% coverage and making the increases in access permanent. In practice, VLOM proved to be insufficient to address the problem of sustainability as communities seldom, if ever, had actual legal ownership, and were ill-prepared to take on the management responsibilities.

By the mid 1990s, the discourse had developed further towards the combination of DRA and community participation. This differed from the VLOM approach in that stronger emphasis is given to communities' demand for services expressed, among other things, through their contribution to investment and operation costs. Greater attention was paid to preparing communities to take on management tasks. As a result, over the last two decades or so, it is the CBM model that has emerged as the leading paradigm for providing water to rural communities. Much effort towards the end of the 1990s went into better understanding the reasons for the success and failure of communities, identifying factors ranging from supply chains to gender, community financial contributions, legal ownership and existence (or lack) of suitable low-cost technologies. Many organisations (particularly NGOs) gradually improved the quality of their project cycles with communities, even though there was an underlying tension between the quality of interventions (both time and resource intensive) and the scalability of approaches.

CBM has undoubtedly brought many benefits, and recent studies indicate that this approach has indeed improved the performance of water supply systems in some cases (Bakalian and Wakeman, 2009). Nevertheless, many communities kept on struggling with sustaining their water supplies, with some succeeding and others failing, giving rise to the notion of 'islands of success' (Davis and Iyer, 2002). The recognition that there are limits to CBM, and that the vast majority of rural communities require some form of on going external support, gained momentum in the early- to mid-2000s, giving rise to a more general focus on post-construction support mechanisms (Lockwood, 2002; Moriarty and Schouten, 2003).

In its simplest form, CBM relies on voluntary water committees to carry out basic day-to-day O&M and administrative tasks to keep the system going and to address minor repairs. This is still a common approach in the smaller communities, typically served by point source supplies, and is evident in the least developed cluster of countries in the study; but is also still present in more remote and isolated rural communities in places such as **India** or **Honduras**. Under this scenario water committees, or CBOs, as they are sometimes referred to, are formally recognised under sector policy and (normally) vested with authority to manage the systems by the relevant authority. However, with exception of the **USA**, CBOs are never the legal owners of the assets they are supposed to manage. Asset ownership typically lies with local government or national government, reflecting the difference between the service authority and service provider functions, with CBOs merely being providers. The concepts of ownership and sense of ownership have been subject to debate in literature on CBM (e.g. Moriarty and Schouten, 2003). Although a sense of ownership has been widely promoted as a basis for CBM, in reality the formal legal ownership of assets by CBOs has often been unclear or never fully assured under national and local laws. However, in view of the separation of the provider and authority function, that could be less important. More important is whether the CBO has a 'sense of being a service provider' or is a service provider in the legal sense of the word. And that is where gaps still exist in the legal frameworks of some countries, and which can undermine the authority and capacity of such committees, such as in the case of **Ethiopia** where they are not yet legalised.

Professionalisation of CBM tends to happen as communities increase in size and tariff-base, and/or where piped networks become the norm, especially in larger rural settlements and small towns. The growth in population and development of more densely populated rural villages are strong drivers for increased levels of service, moving up the ladder from basic point sources, including handpumps, to reticulated systems with street or household connections. This trend towards professionalisation is normally accompanied by various opportunities for private sector involvement, either through sub-contracting specialist functions to individuals, such as paying to have a plumber or bill collector. Finally, in the larger and more complex systems, such as multi-village schemes serving rural growth centres, professional service providers are often contracted to carry out some or all of the operational functions. Considerations of market viability and economy of scale are crucial for the effective take up of such roles. This graduation of CBM options is well-illustrated in the case of **Ghana** which identifies three broad sub-models as shown in Table 7. In **Ghana** the DA has the power to designate, and delegate authority to WSDBs, which hold the system

Table 7: Ghana: Formally recognised sub-models of CBM

Population size	System	Management model
< 2,000	Point source	WATSAN committee
2,000 - 5,000	Non-mechanised systems (e.g. gravity-fed water schemes)	WSDB supported by skilled artisans from within the community, whose services may be procured when necessary on a retainer basis (indicated as 'option 1' in the CWSA Small Towns O&M Guidelines).
5,001 - 10,000	Simple boreholes, gravity fed or slow sand filtration based piped systems	WSDB with certified/reputable firm to carry out specialised functions as and when needed (indicated as 'option 2' in the CWSA Small Towns O&M Guidelines) or, preferably, WSDB with contract with a firm or firms to perform specialised functions on a periodic basis (indicated as 'option 3' in the CWSA Small Towns O&M Guidelines).
> 10,000	Communities served with complex water supply systems	WSDB and a contracted firm (private operator) to completely operate and maintain the water supply system (indicated as 'option 4' in the CWSA Small Towns O&M Guidelines).

Source: IRC/Aguaconsult, 2011

in trust, and manage service delivery on behalf of the DA. In turn, the WSDBs can further out-source specific functions to individuals and private companies, depending on the size and complexity of the systems. In some countries this differentiation starts to be recognised and formalised, based on characteristics such as type of system or village size. For example, in **Thailand** there are four different and increasingly professionalised forms of CBM stipulated in **Tambon** local government policy.

One of the major success stories in CBM, and the only real example from the case study countries where this has been adopted at scale, is the WASMO[3] from the state of Gujarat in **India**. WASMO is active in every district in Gujarat, and provides rural water services in about 15,000 of the state's 18,000 rural communities, representing some 26.5 million people. In its early days WASMO was driven by Dutch development assistance when the model emerged to take over from the centralised Gujarat Water and Sewerage Board (GWSSB), which had exclusive responsibility for rural water up until 2002. The main highlights and factors for success of the WASMO experience can be summarised as follows:

- WASMO is not only scaled up in terms of coverage, it also functions 'at scale' meaning that it is a self-sustaining model that has support structures, implementing systems and financing mechanisms which can be replicated. There are still challenges, and one of the main difficulties is to deal with district bureaucracies and the special problems of the tribal areas.
- Under WASMO community management has been truly formalised by modifying bye-laws that made the water committees (*Pani Samitis*) a legal part of local government (*Gram Panchayats*), thereby giving them a full legal status.
- The role of the *Pani Samitis* is clearly defined; the physical assets of the system are still owned by the state, but this relationship is governed by a tripartite agreement signed between WASMO, the local *Gram Panchayat* and the *Pani Samiti*.
- One of the critical success factors has been the political dimension, with a deliberate plan to engage with politicians and senior civil servants from the outset, and to have them informed and supportive all the way through.

Even though progress is evident on the ground, there has not yet been a full-scale evaluation of WASMO which would provide hard evidence and metrics around whether this approach leads to better quality of service and improved sustainability.

As CBM has become more complex and professionalised there is increasing out-sourcing of technical functions and, in some more isolated cases, the entire management and operation is contracted to private operators. This often blurs the distinction between out-sourcing and true *delegation*. Delegated management, as a concept borrowed from urban utility experiences, refers to all forms of contractual relationships between an asset owner and an operator. The key aspect of formal delegation is that this can only be done by the asset holder; only to a legally recognised entity (which is usually a private company, but also could be an association or community committee); and is subject to a service agreement setting out the terms of the contract and services expected. In this sense, much of what goes on under more professionalised forms of CBM is not strictly speaking delegation, but rather sub-contracting, as it is not done under service agreements, and without CBOs as asset owners. For example, in some states in **India** (e.g. Maharashtra) and in **Sri Lanka** it is common for the out-sourcing of functions to be made to local NGOs or CBOs which act as private contractors. More conventional cases of delegated management for rural water systems are presented in the section 'Beyond CBM', p81.

The transition of CBM from basic VLOM to more professionalised forms over the past 20 to 30 years appears to be driven by a number of factors. The

> **Box 3:** Colombia's entrepreneutrial culture programme
>
> A review of progress on sector reform in 1998, six years after its start with the Law on Public Services in 1992, revealed that a high percentage of small municipalities had failed to complete the procedures to legalise the water utilities. It also showed that infrastructure was still being built without having a solid service provider responsible for operating that infrastructure. This was one of the key factors for creating a programme of assistance initially called *Programa de Cultura Empresarial*, best translated as entrepreneurial or business culture. In 2003, the programme was renamed into the programme for Strengthening and Technical Support to Small Municipalities, but kept its main premises.
>
> The Programme has three objectives: 1) to establish and/or legalise *community-based* water service providers in rural areas and small municipalities; 2) to support the development of a business structure among these service providers; and 3) to improve service provision indicators among the providers that participate in the programme. One of the principles of the programme is the recognition that CBM is the main and most relevant service provision option in rural areas, but CBOs need to operate as formal service providers, operating under basic business and entrepreneurial principles, even while they continue operating as non-profit organisations. Initially, a big effort of the programme was on dissemination of the legal and institutional framework and requirements among municipalities and operators. Later on, more practical tools were provided that allowed operators to become more professional, including the provision of training material on issues such as billing and tariff collection, book-keeping and financial management, operation and maintenance and customer relations. Operators which have made progress on certain criteria also received a free license for billing software. An evaluation by Tamayo and García (2006) found the contents of the programme were useful in structuring more professional operators. However, it also identified gaps including the limited emphasis on community ownership of the service they would receive.
>
> **Source:** Rojas, et al., 2011

key one seems to be demand to move up the service ladder to systems that can provide higher service levels, particularly more complex piped systems with household connections. In turn, one of the unseen drivers for this demand-driven move to better services – and therefore more professionalised management and operation – appears to be urbanisation and (re)migration. In some middle income countries such as **Sri Lanka** the growing shortage of labour in rural areas means that there are fewer communities willing to perform 'free' services, including technical tasks of the water committee (e.g. monitoring water quality, checking for leaks, etc.). Another key driver is a simple change in philosophy among policy makers that the provision of a key public service such as rural water supply also requires professional service providers. This has been evident in **Colombia**, where the *Programa de Cultura Empresarial* (entrepreneurial culture programme) has focused on the promotion of good management and business principles, while retaining the not-for-profit status of community management (see Box 3 for a summary). In fact it is already telling that in **Colombia** the term Community-Based Service Providers is used to describe the CBOs running a water supply system, whereas in most other Latin American countries 'water committee' or 'water board' is the most commonly used term. This difference in terminology reflects the conceptualisation of more professionalised service providers. It also reflects the fact that professionalisation should not only be understood in terms of hiring professional staff; it also entails attempts to operate at high standards and run a service provider in a professional manner.

Regardless of the underlying drivers, some of the key defining characteristics of this professionalisation trend within CBM include:

- The separation of service provision functions from operations functions – wherein communities through their elected representatives (either in local government or in CBOs such as Water Boards) retain the ultimate management and decision-making power, but are able to separate out specific tasks, or retain all of the operation and administration of a system, and delegate this to individual entrepreneurs or local companies.
- A change in philosophy from volunteerism towards professionalised approaches which may include a business model or management culture.
- A strengthening of the capacity of service providers by improving performance based management and adoption of good business practices.

Although post-construction support has been recognised as a critical aspect of successful CBM, it is interesting to note that in most cases this is not yet seen – or planned for – as an integral part of CBM. Even where functions for such support have been allocated to local government and the options exist, such as in **South Africa** and **Colombia**, this is not adopted in a systematic way. Rather, it is left to individual local or regional authorities to give shape to this function. Only in **Honduras** has the question of post-construction support been addressed in a systematic way as part of the CBM model. Post-construction support is addressed in more depth on page 103.

Beyond CBM: private sector management arrangements and rural utilities

As noted above, although there are elements of contracting out within CBM, formal delegation arrangements are a relatively recent phenomenon for management of water provision in rural areas. Nonetheless, bringing in local private operators with more specialist or professional skills can be a way of improving services and increasing efficiencies over the more conventional CBM approaches. With population growth, urbanisation and migration trends all increasing, there is a growing number of larger rural communities, growth centres and small towns, all of which present more complex water supply needs and demands for higher service levels. This, in turn, brings a greater likelihood for professionalisation of services as tariff bases grow, and there is a more skilled workforce available in such growth centres. These contexts are well beyond the scope of conventional CBM arrangements more common in small, low-density rural villages, and require a step-change in management arrangements.

The case studies indicate that these types of private operator arrangements are emerging as a small, but nonetheless significant option, and have been adopted in a number of countries, including **Benin**, **Burkina Faso**, **Colombia**, **Ghana**, **South Africa** and the **USA**. There is documented evidence from other countries and regions, particularly in French-speaking West Africa and Rwanda.

Since 2000 in **Uganda** the DWD has been piloting and expanding a model for delegated O&M of Small Town water supply systems based on local private sector engagement. These O&M contracts have been typically short in nature (a three-year rolling contract arrangement), and place minimal requirements for capital investment or system expansion on the operator. The district local government acts as a water authority, and signs and supervises contracts with the private operators, with DWD playing a technical advisory and support role through its regional TSUs. In 2005 the DWD started work with the Global Partnership on Output-based Aid to make the contracting conditions more attractive to local private sector operators; today 72 RGC systems are run by private operators, representing 8.5% of the national total (Azuba, Mugabi and Mumssen, 2010).

Another country which shows a relatively high degree of development of delegated approaches to managing service delivery is **Benin** (as it is also in **Senegal**, where there has been a recent policy shift in favour of fully delegated management approaches [AGUASAN, 2008]). It has established a range of models for delegation directly between the commune and private operators (see Box 4). The advantage of this approach is that it allows delegation of management contracts for a larger geographical area, i.e. for more than just one community. This allows for economies of scale. Delegated contracts for rural piped networks ensure that part of the revenue goes to a fund for CapManEx and network extensions and also to the commune's own general budget. The example

> **Box 4:** Different approaches to delegation in Benin
>
> Service Delivery Models in Benin differ between different types of technology. The most common model is still the basic CBM approach with a water user association (service provider) acting as the operator. However this is changing as professionalisation is pursued and alternative management structures are put in place. For 'simple' technology such as handpumps there a number of options of delegation by the commune
>
> - Delegation to a community representative
> - Delegation to a private operator
> - Delegation of many similar systems to one local operator
> - Delegation to one operator of different types of system (e.g. handpumps and piped networks) within a geographic area
>
> For the more complex piped networks or mechanised boreholes there are other more complex models recognised under the legislation, but some of these are not common:
>
> - Delegation to a private operator
> - Tripartite contract involving the commune, water user association and a private operator
> - Delegation of production to a private operator and distribution to a water user association
> - Delegation to a water user association
>
> Overall in Benin the delegation process is open tendering with positive discrimination for local entrepreneurs to encourage local private sector development, where national private operators from the cities are excluded.
>
> **Source:** Adjinacou, 2011

from Benin illustrates the possibility for formal delegation from commune to water user associations and private operators.

Another example of private sector involvement was the Management Reform of Rural and Small Town Water Supplies Programme in **Burkina Faso**. After disappointing results from running a CBM model for many years, a private operator model was tested. The government began a new strategy under which private firms were awarded a handpump maintenance or a handpump installation and maintenance contract in one or more communes. Smaller firms were awarded the former type of contract covering one or two neighbouring communes, and larger firms got the latter, covering several communes in a region. For rural piped schemes, the commune contracts an operator to manage each system. The promotion of these delegated contracts to private entities has met with some success, especially in achieving greater economies of scale. One constraining factor in this approach is that commune staff often do not have the capacity to understand and monitor these more complex agreements.

In the above cases the private operator is contracted only to manage the service. But in other cases the contract goes one step further and the private contractor would also be responsible for the construction of the system. The cases from the study countries and others in the general literature highlight the growing involvement of local, small-scale private operators as an alternative model to CBM for improved water supply services in small rural towns and rural settlements. The concept of these PPPs is gaining ground and is starting to be documented (see Kleemeier, 2008; WSP, 2010). In generic terms such PPPs rely on a central tripartite arrangement between the contracting authority (the asset holder, which under decentralisation is often the local government), the operator, and some type of regulatory body. Given that in most countries the formal regulatory framework for the rural sector is either non-existent or extremely weak, this function is often still played by a centralised ministry and its deconcentrated offices, as is the case in **Uganda** where this role is still played by the DWD. In addition to these three main actors, PPPs also normally involve some form of support agency to help guide and monitor the contractual relations between the asset holder and the operator. In some West African countries and in **South Africa** this role has been taken up by private sector support agencies (also termed 'business development service providers' and SSAs respectively).

Self-supply

Self-supply is, and historically has been, an important form of managing domestic water. It refers to a situation in which individual households (or sometimes even a group of neighbours) invest in gradually improving their own service (Sutton, 2007), and where the O&M is also done by the household themselves. Self-supply can fill the gap where public or formal private sector-led approaches do not reach. This is especially the case in scattered rural communities and where water sources are easily available, for example, in many parts of **Bangladesh** or **Zimbabwe** where water tables are very high. But, it also happens in small towns or peri-urban settings, where better-off families invest in their own borehole and gradually extend services to neighbours as seen, for example, in **Benin** and **Ghana**. Of course, wherever there is no (adequate) service provided by a public, private (formal or informal) or community entity – everyone who is not served in effect comes under the self-supply model and it is therefore, *de facto*, an important approach to consider. One of the main challenges to date has been that self-supply is not easily recognisable or quantifiable, and therefore usually does not qualify as a management option with formal benchmarks.

One of the main advantages of self-supply is that household-owned systems tend to be better maintained than communal systems. In **Zimbabwe** sector statistics show that family wells are generally better maintained than communal systems; non-functionality rate for family wells was 12% in 2004, while for boreholes with bush pumps it was 30% (UNICEF/NAC, 2004).

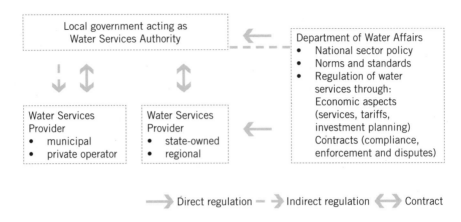

Source: Based on de la Harpe, 2011b

■ ■ ■ **Figure 7:** Arrangements for letting and regulating delegated contracts by the WSAs in South Africa

Reasons include the fact that households are more likely to maintain a water system that is also legally their own and that does not require relatively complex CBM arrangements. In addition, the costs for maintaining such systems tend to be lower.

This does not mean, however, that there is no role for government agencies or NGOs in self-supply. They may promote self-supply as an appropriate measure both at the low end of the coverage scale (where accelerating access is required), as well as at the very other extreme, where coverage levels are very near

---

■ ■ ■ **Box 5:** Promoting rainwater harvesting to support self-supply approach in Thailand

Rainwater harvesting has traditionally been one of the most important sources of water for rural households. In the mid 1980s, when the Thai Government (RTG) started decentralised approaches to rural water supply, it focused in on three low-cost technologies, including rainwater jars for drinking water supply, alongside shallow wells for other domestic uses. This was before piped village water systems became the driving model for water delivery.

Rainwater harvesting programmes have since then been undertaken by village committees, government and NGOs, to support household investments. This led to an increase of rainwater harvesting as a source of domestic supply from 24% of the rural population in the late 1980s to 37% in 2005. The combination of self-supply with supporting programmes helped achieve economies of scale in the supply chain and created competition in the private sector. This in turn has pushed prices down of the jars to around US$400 for a jar of 11 m^3, which can provide water security for a six person household all year round.

Source: Juntopas and Naruchaikusol, 2011

complete, but where reaching the remaining percentage of unserved is often the most challenging and costly, as these people tend to live in the most dispersed rural areas. Such promotion can take many forms, including the provision of awareness raising and technical assistance on practical ways of installing household systems, to providing subsidies to the installations.

A good example of this is **Thailand's** approach to promoting rainwater harvesting (see Box 5). Outside the study countries, there are other well-known cases of government support to self-supply. In **Zimbabwe**, since the early 1990s, government and donor programmes have supported the protection and upgrading of shallow wells. By 2006 over 120,000 family wells had been up-graded, serving more than 1.5 million people at minimal subsidy by donors or the government (US$3-5 per capita) (Sutton, 2007). **Ethiopia** has started a similar programme. This also shows that where government provides some support, it is able to leverage households' own investment.

However, such programmes of support are not common. As a result, self-supply remains largely invisible in sector indicators and, therefore, its relative importance to the sector is not fully understood nor appreciated. However, this trend is starting to change, and as seen in Table 6 (page 74), 8 out the 13 case study countries recognise self-supply in sector policy (**Ethiopia**, **India**, **Thailand**, **Uganda** and the **USA**). In the USA, where self-supply is regulated much more stringently than other models, some 14.5 million people are using privately financed and operated systems. **Honduras** used to have a programme of support to self-supply as well, but this programme stopped and, as a result, there is no insight into what percentage of households may be being served through self-supply. Formally recognising self-supply makes it possible to see the investments made by people themselves, and also to direct more limited support to improve these self-help services, which is often needed to ensure the improvement of sources.

Discussion: defining Service Delivery Models as part of the Service Delivery Approach

Based on the findings of the various country studies it is possible to identify a categorisation of SDMs, which comprises four main options, namely: CBM, direct public sector provision, private operators, and self-supply. But it is clear that in reality there are a number of variants within most of these categories, reflecting different degrees of system complexity and levels of professionalisation. The most important variants in management arrangements under these different SDMs are given in Table 8.

While the cases studies do not indicate any right or wrong SDMs, what is clear is that there has been a generalised trend away from the more voluntary arrangements of CBM towards professionalisation, or what some are now terming 'community management plus'. Some of the founding principles of CBM, such as community cohesion and common participation for the greater good and informal accountability to a water committee, have been seen

Table 8: Service Delivery Models and variants

Principal SDMs	Main variants in management arrangements
CBM	• Small-scale systems directly managed and operated by water committees • Larger and more complex systems managed by water committees with individual operational functions out-sourced to private individuals or small companies • Larger and more complex systems managed by water committees with all O&M out-sourced to private sector operators • Associations of water committees providing economies of scale for certain O&M functions
Public sector operators	• Municipal water company providing management services to rural communities • Associations of municipal companies providing management services to rural communities • Local, regional or national public utilities providing management services to rural communities, including maintenance contracts
Private sector operators	• PPPs, with private sector operators to maintain and manage larger systems under contract • Formal private operators working under licence • Informal private sector providers • NGOs and CBOs
Self-supply	• Individual households • Clusters of or neighbouring households

to be problematic over the last 20 years or so. While it can be argued that the reverse is also true – that two-thirds of community-managed systems still function – tolerance of this level of failure would not be contemplated in most spheres of the public service. This ideal has been undermined by the lack of formalisation of these arrangements within broader local government bye-laws and national legislation and policy, the absence of clear contracting, lack of legal standing of the committees, and the lack of professional capacity in certain aspects of running and managing systems. A second and related trend illustrated by the studies is that there is an emerging but growing role for small-scale private operators to improve service provision for rural populations. Although these approaches only account for a small proportion of management models in some contexts at present (i.e. in **Ghana** this is estimated at only about 4% of the total [IRC/Aguaconsult, 2011]), in other parts of the world it is growing rapidly. For example, in a number of countries in West Africa, PPP arrangements are now in place for about a quarter of all piped schemes (both rural and urban), and this proportion is expected to rise quickly in the coming years in a range of countries including **Burkina Faso, Mali, Mauritania, Niger** and **Rwanda** (WSP, 2010).

The demands on management and administration of a water supply system are, in part, a function of the technology employed. Household level supplies, including rainwater harvesting, present the simplest challenges in terms of management, but even here there is the need for some form of follow-up to address issues around safe storage, water quality and hygienic use. Point source technologies, including both handpumps on wells or boreholes and capped springs, require less management capacity and sophistication than large reticulated systems with household level connections. The difference between gravity-fed systems and any requirement for a power-source to pump water, whether this is electrical, fossil fuel or solar, again adds elements of complexity (and cost) to maintaining the system and delivering water.

In recognising different service provision models, it is useful to differentiate these in their feasibility according to settlement type, which in turn reflects both changes in population growth and settlement patterns, meaning we are seeing more and higher-density rural growth centres. Figure 8 maps the different formal models found in the case studies against these different demographic settings. This reflects that highly dispersed communities tend to rely on voluntary management or self-supply, and more concentrated villages and rural growth centres and small towns have a larger spectrum of options available. This spectrum highlights the immense difference in demands and solutions between low-density rural villages and hamlets and emerging small towns or rural growth centres. Opportunities for increasingly professionalised service delivery with the corresponding improved management capacities and revenue collection are clearly more abundant in the latter case.

Using a broader service area (like the *Communes* in **Benin** which may contain a mix of settlement types) is one way of providing economies of scale,

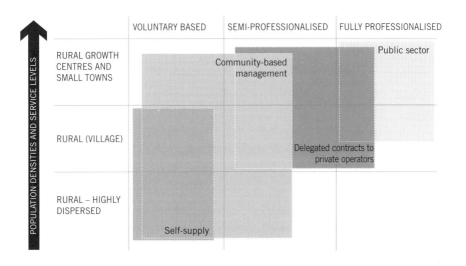

Figure 8: Service Delivery Models and types of settlement

thereby facilitating delegated options. Various countries also differentiate the type of service provision model according to the type of technology (which is often closely associated with the settlement type). Both **Burkina Faso** and **Benin** are examples of that; in the latter case different delegated models are promoted for simple point sources and more complex small piped systems. The more actual options are elaborated in sector policies and guidelines, the greater the likelihood of finding a good fit between a model and a service area. The presence of different models may even lead to further innovation in service provision.

The adoption of an SDA calls for identifying pathways for professionalising SDMs. It is noteworthy that there are different preferences for professionalisation in different geographic contexts. Countries such as **Ethiopia** and **Mozambique**, which are still struggling with very low levels of coverage, have much more basic forms of community management and fewer other formally sanctioned models. But countries like **Benin**, **Burkina Faso** and **Ghana** employ a wider range of delegated contractor options as one of the pathways to professionalisation. In the Latin American study countries the delegated model is not much favoured, but rather there has been an emphasis on professionalising the existing community-management structures through an institutionalised system of post-construction support.

Service authority functions

The different management options described in the previous section require different forms of support from the intermediate level, normally local government which, as part of sector reform and broader decentralisation processes, have become service authorities. (Service authorities implies being responsible for *guaranteeing* access to services and being the asset owners – though in some countries that responsibility remains with national government.) Service authority functions, are critical to supporting service delivery and are established by national policy, legislation and, to some extent, local bye-laws. The service authority functions cut across a range of issues from planning and implementation, to monitoring, regulation and post-construction support.

Planning and implementation of water services

The first set of authority functions relates to planning and decision-making on water interventions, as well as implementation, although this latter task may be contracted out. Here we look at the planning and implementation approaches followed. Specific attention is given to the link between water services and water resources planning.

Strategic planning

Different planning instruments are to be found in the case studies. In some countries local governments are expected to develop strategic plans for prioritising water interventions, either as stand-alone 'municipal water plans' or as a 'water chapter' of broader local development plans as is the case in **Ghana** and **Burkina Faso**. However, the experiences with strategic planning at local level are mixed. Experiences in **Uganda**, for example, show that planning tends to focus mainly on implementation of new water services and some limited rehabilitation, but tends to ignore the need for replacement of existing assets or planning for support to those already functioning. A similar experience was found with the WSDP in **South Africa**, which often was little more than long wish-lists of projects, with inadequate planning for asset replacement. One reason for that might be that the local governments see themselves often as an implementer of infrastructure only, and not yet as owner of assets for which they bear ultimate responsibility. It may be complicated when systems are developed by NGOs or projects, of which the local government has no knowledge and over which it might be reluctant to take on formal ownership, let alone include them in their asset management plans. A second reason relates to the poor quality of the process for developing such plans. Often these have been technical exercises, driven by local consultants, with limited involvement of local politicians, local government officials or consumers; this is often due to the limited capacity of local governments.

In addition to strategic plans, which normally cover a longer time frame (i.e. three to five years), local governments often have to make annual plans, linked to annual government budgeting and disbursement cycles, as is the case of the WSDP in **South Africa** which covers a five year time span within an annual planning context. Annual planning – and, more specifically, annual financial disbursements – can often be problematic when there are delays and lag times in the release of funding. When disbursements are only made towards the end of a financial year this can often lead to rushed spending on implementation and poor quality, as in the case of some districts seen in **Uganda**.

Planning is not done at local government level in all cases, rather in certain countries this has been elevated to higher levels such as the province, amongst others in **Benin**, **Burkina Faso** and **Mozambique**. This leaves the local government mainly in a role of channelling information between communities and provincial authorities. In other countries authorities at higher hierarchical administrative levels play a role in planning processes. For example, the TSUs in **Uganda** play a role in quality and consolidation of annual local government plans. **Colombia**, as also described in the section on p66, is moving towards making plans at departmental level, to which all local authorities contribute. This 'local centralisation', as it is dubbed by OECD (2010), is to allow both economies of scale in the use of resources (such as engineering capacity), and the planning for costs such as direct support costs.

Implementation approaches

In addition to strategic and annual planning, which typically cover an entire area of local government or provincial jurisdiction, there is a third type of planning which covers the implementation of capital projects. We refer to this type of planning as the project implementation approach, which entails all procedures and mechanisms for the capital intensive period in the cycle of a water system. Historically, many programmes have focussed almost exclusively on this as a discrete planning cycle, which is often completely divorced from the broader service delivery cycle as shown in Figure 1 (p18).

Many lessons have been learnt about the importance of the quality of implementation approaches and interventions for sustainability. Key factors identified include the degree of participation of communities of users in decision-making (Harvey and Reed, 2006; van Wijk-Sijbesma, 2001), gender, technology selection processes, and the development of CBOs. Failure to address these factors in the intervention – particularly in new system construction – often leads to rapidly emerging sustainability problems. Although the opposite is not necessarily true; having good quality intervention approaches is not a guarantee for sustainability.

Many countries have tried to include these factors in a standardised manner in implementation guidelines. Countries like **Uganda**, **South Africa** and **Mozambique** all have project guidelines or implementation manuals which describe in detail the steps in the implementation of water systems, as well as providing tools. In these countries there is a strong drive to ensure that most, if not all, agencies operating in the sector follow these manuals, with the aim of ensuring that intervention processes meet certain standards, both for the technical quality of the works, as well as for the software interventions For example, in **South Africa**, practitioners working on the process of community mobilisation are required to follow a standard training in what is called the 'Institutional and Social Development (ISD) Package'. Another aim of such guidelines, as with the case of **Ghana**, is to harmonise approaches and reduce confusion when different development partners implement projects. Not all countries have such standardised intervention processes. For example, in **Honduras** there is no standard project implementation manual. Despite this, most agencies follow the Escuelas y Casas Saludables (ESCASAL) (Healthy Schools and Homes) approach to hygiene education and demand creation. The same applies in **Colombia** where different local and provincial government agencies have their own implementation manuals, albeit with similar approaches.

It is not always easy to ensure that all agencies follow standardised implementation approaches that allow for a minimum level of standards to be met. NGOs in particular are less likely to follow a country's standard implementation manuals, as many often have their own organisational approaches. This has led, for example, the Uganda Water and Sanitation NGO Network (UWASNET) in **Uganda** to encourage member NGOs to adopt the formal Government of Uganda implementation manual and procedures. One way to improve approaches is to hand over even more responsibility for project implementation to communities. In **India**, **Honduras** (the community executed projects – known as PECs) and **Ethiopia** (as part of the Community Development Fund [CDF]), communities are not only responsible for the eventual provision of the service, but also for managing the entire implementation cycle, including procurement of materials and contracting. This has reportedly strengthened ownership over the service, improved transparency in the use of funds, and improved the quality of the systems. At the same time, this approach is also more time consuming as the community needs additional training and support to fulfil its role in implementation adequately.

In the many instances where an SDA has not been applied or planned for from the outset, there is often an abrupt disconnect between the capital intensive phase – the new 'project' – and the subsequent transfer of responsibility to communities under CBM arrangements. In some cases there may be a few months of mentoring, or even more structured post-construction support programmes (on p103). But even in those cases it often means a transition for the community from one agency with whom they engage in the construction phase (e.g. government, donor or NGO) to another agency providing post-construction support, if they are lucky enough to have this in place. For systems managed by private operators, individual user households and the community may be less aware of this transition.

In addition, the separation of roles and responsibilities, discussed in section 'Institutional arrangements for rural water supply' p65, has also led to the separation of the implementing role from the support or regulating role. That, in itself, need not be a problem as long as the transition in the service delivery cycle is clear. In that sense, Figure 9 provides an interesting illustration of how **South Africa** conceptualises the changing roles for different agencies at different moments in the service delivery life-cycle.

Water resources planning

A specific area of planning is water resources planning. Most of the planning instruments described above focus specifically on water supply services, but because this is dependent on broader water resources availability there is the need to assess how water resources are incorporated.

In general, rural water supply is not a big user at river basin level. Globally, the domestic water supply sector represents approximately 13% of all water abstractions (FAO, 2010), and rural water supply would be a small part of that total figure. However, rural water supply can be affected by increasing consumption by other water uses such as irrigation or urban water supply. The case studies show that this is only an issue in countries that are experiencing what is termed physical water scarcity, i.e. a situation where there is not enough water to meet all demands. Among the case studies only parts of **India, South Africa** and the **USA** face such a situation (Comprehensive Assessment of Water Management in Agriculture, 2007). All other countries face

	PHASE 1 Policy	**PHASE 2** Planning	**PHASE 3** Implementation	**PHASE 4** Service provision
National sphere	Develop municipal infrastructure policy and set standards for delivery systems	Develop framework for National Spatial Development Perspective (NSDP)	Municipal infrastructure programme management, collaboration, mobilise support and monitoring	Regulatory oversight of systems, procedures
National water sector sphere (DWA)	Develop water sector policies, norms and standards	Macro water sector planning (part of SWAp)	Monitor implementation of norms and standards and collaboration around support	Regulatory oversight of water services norms and standards
Provincial sphere (DWA provincial office)		Provincial Water Services Plan (links to SWAp)	Monitor the implementation of sector policy, norms and standards and provide sector support	Service provision support
Municipal level	Service provision policies and bye-laws	Integrated Development Plan (IDP)	Infrastructure delivery systems put in place and project management	Municipal regulation over service provision
Municipal water sector level (WSA)	Water policies for free basic services	District / local water services development plan	Technical department (e.g. water, roads, etc. oversight of project implementation) or infrastructure unit	Service provider agreements
Water Services Provider	N/A Not responsible for service provision policy	WSP Business Plan (plan for *providing* services in service delivery area)	May be contracted by WSA to implement capital projects	Operations, maintenance, customer care, revenue collection, etc. as per service provider agreement / contract
Project level		Project Pre-Feasibility and Feasibility Studies and business plans Design Tender Construction Commission	Project cycle – implement technical norms and standards	Incorporated into existing service provision arrangement, or new WSP established as part of project cycle

Source: de la Harpe, 2006

Figure 9: Sevice delivery life-cycle in South Africa

no physical water scarcity, but rather economic water scarcity, i.e. human, financial and institutional capacity are too limited to harness water resources, even though they may be naturally abundant. In practice, this does not preclude local scarcity and conflicts between users of rural water supply systems and other uses in such countries.

It is exactly for this reason that the (rural) water supply sector traditionally has engaged only to a limited extent with water resources institutions, as the latter initially have focused on much larger-scale river basin scarcity (see also Moriarty, Butterworth and Batchelor, 2004; and Smits and Butterworth, 2006 for further discussion on this). Rural water supply agencies often have a narrow conception of water resources management, focusing on activities such as protection of springs and streams in upper catchments, as reported in the cases of **Colombia** and **Honduras**, for example. But they tend to ignore more complex relations and competition with other water uses. However, in view of the increasing scarcity and recognition of the localised nature of competition between rural water supply and other uses, a range of efforts are undertaken to strengthen this linkage:

- Using water resources data as technical input into rural water supply planning. This is the traditional approach of assessing water availability as a basis for technical decision making on, for example, the siting of a borehole or the design of surface water intakes. Most of the countries in the study follow this approach. **Uganda**, for example, has undertaken efforts over the past years to make groundwater maps available to local authorities so they can take such data into account in planning. This approach works relatively well, even in the absence of local competition for water resources.
- Regulating rural water use through licensing. This approach recognises that local water resources may be limited, and that these can be regulated through 'technical' control, via abstraction licenses. This approach is used, for example, in **Colombia**. The main disadvantage of this approach is that the administrative transaction costs are very high: every small water system needs to apply. Monitoring and control costs would also be high. In reality, it is often found that rural water supply systems do not apply for such licenses to avoid these costs, or are not monitored for licensing.
- Licensing, but with exceptions for small users. This approach is similar to the previous one, but with the important difference that users that use less than a given amount do not need to apply for a license. This approach is followed, for example, in **South Africa**. The advantage of this approach is that only the big users need to apply and can be regulated, and small users would be indirectly protected from overuse. This also brings down administrative transaction costs.
- Integrated planning between water supply services and water resources. Under this approach water resources are not only taken as a technical input into water supply planning, but also account is taken of existing

water allocations (as defined under the licenses) and, where needed, these are further negotiated. Of the case studies, **South Africa** is the country that probably has the most advanced institutional relationship and planning instruments. WSAs need to align their WSDPs' plans with the constraints placed by the relevant Catchment Management Agency. There is both an institutional and planning relationship between the WSA and Catchment Management Agency.

- Protecting and increasing supply. This approach is fundamentally different from the other approaches, as it seeks a way of reducing competition with others uses by protecting or even increasing supply for rural water. This approach is based on protecting rural water sources such as springs and streams. But it builds on it, by promoting the harnessing of additional or alternative water sources. For example, in **India**, emphasis is placed on providing communities and households with one or two alternative sources beyond the main source, e.g. through rainwater harvesting or use of traditional sources. The rationale behind this is that secure access to one source only is unlikely in the context of high competition, so a better solution may be to diversify access to a number of sources, so that if one source fails, there is a back-up option. Obviously, this has implications for costs of source development.

Accountability, regulation, oversight and enforcement

A key aspect of the adoption of a more service-oriented approach is the establishment of accountability mechanisms; in other words, the ways in which consumers can hold service providers to account for the actual service they receive. For the rural water sector it is important to make a clear distinction between what is meant by accountability and the related functions of monitoring, oversight and enforcement. The most formal manifestation of accountability mechanisms is *regulation*, which is typically found in the urban sector and derives from developed country contexts, often relying on one regulatory body (e.g. the Water Services Regulation Authority [Ofwat] in England and Wales). The main purpose of regulation is to provide a set of rules, norms, monitoring and enforcement processes that ensure service providers meet nationally set guidelines and standards. The regulatory body is also responsible to ensure a level playing field for users who may otherwise suffer in a monopolistic situation or from price setting between a number of providers (Water Partnership Programme/AfDB, 2009).

Few developing countries have an independent and effective regulatory body for urban water utilities, much less for rural services. Nonetheless, as rural operators become more professionalised or where local private operators take up an increased share of the market, regulation and holding such providers to account will be an increasingly important function. As well as establishing a regulatory body, regulation itself also entails many actors and processes, including sector ministries, asset holders and service authorities, consumer

Indirect or 'long arm' of accountability going via government entity

GOVERNMENT MINISTRY OR REGULATORY AUTHORITY: can apply accountability measures or formally regulate service providers

CONSUMERS: households and communities paying tariffs to receive service from providers

Direct or 'short arm' of accountability between users and providers

SERVICE PROVIDERS: Community-based or private entities with responsibility to operate and maintain systems – formal or informal contracts

Source: Adapted from World Bank, 2004b

Figure 10: The framework of accountability relationships

groups, independent monitoring agents (sometimes contracted in) and the operators themselves in cases where elements of self regulation are incorporated. Norms and standards are largely set through policy and legislation at central level, but regulation and oversight also depends upon bye-laws, locally let contracts and the relationships between actors working at the district or municipality (Water Partnership Programme/AfDB, 2009). Because in the rural context these functions are almost always put into place by local actors, regulation and accountability are included as part of the service authority functions in this book.

A key concept when considering accountability in the rural context is the differentiation between the so-called 'short' and 'long' arms of accountability (World Bank, 2004b). In this case, the former refers to a direct accountability relationship between consumers and service providers, and the latter to the mechanisms through which consumers hold public policy makers to account, who in turn regulate or hold the service providers to account. However, as stated by the World Bank (2004b) water supply services are not solely provided through market transactions, but through government taking (partial) responsibility and, therefore, there should also be a form of indirect accountability. In reality these two forms of accountability exist in parallel as they both tend to be vulnerable, and the breakdown of part of the relationship can result in poor service delivery (World Bank, 2004). A formal regulatory framework is one way of improving this type of indirect accountability.

A first finding from the study countries is that for CBM of rural water supply, much emphasis is placed on the establishment of the short arm of direct accountability relations between consumers and their respective water committees, as service providers. This often stems from the typical project cycle where, after completion, the entire service delivery responsibility is placed on the community: the water committee provides water to consumers who pay a tariff to cover costs. Consumers in turn should hold the committee to account for service performance. There is ample evidence that this form of accountability is very vulnerable: there is a high risk of ending up in a vicious circle of poor service delivery, non-payment of tariffs by unhappy consumers and an even further deterioration of services. In several cases the CBM approach is based on establishing a management entity (the 'committee' or 'board') which is not even recognised under formal water sector policy and, in many cases even where it is, these same entities lack any type of legal standing, which makes it impossible for them to enter into formal contracts.

In cases where CBM is legally recognised, one first step in improving accountability is strengthening the direct (or short arm) relationship between consumers and providers. This is typically done during project implementation capacity building where rights and obligations of consumers and service provider are highlighted. It is also part of various post-construction support mechanisms, where these exist. For example, in **Honduras**, where the *Técnicos de Operación y Mantenimiento* (TOMs) (Operation and Maintenance Technicians) of SANAA and other technicians have recently added a module on accountability for their training to water committees. In **Colombia**, part of the package to establish a business culture among rural operators is the development of customer relationship mechanisms. In other contexts – including rural, small towns and urban water supply – positive experiences have been obtained in the use of tools such as citizen score cards, through which consumers can hold service providers to account (Ryan, 2006). However, most of these efforts are limited to places where post-construction support mechanisms exist and which are consistently supported, often with donor funding.

The studies illustrate a number of different examples of how the indirect form of accountability has been established, including regulatory functions, even in cases where there are no formal sector regulators in place. The most basic mechanism is one where there is general water supply regulation established at national level but where local government, as service authority, has an oversight and enforcement role. This is, for example, the case in **Benin**. The main disadvantage of this approach is the risk that the service authority might take a reactive position, i.e. it will only check on a service provider where a problem arises, so as to avoid incurring expenses on proactive monitoring. In many cases, service authorities have limited capacity to monitor during the service delivery phase.

Formal contractual agreements between service providers and service authorities (as asset holders) form a second arrangement modality for regulation, in which a contract specifies the services that need to be provided and within which conditions. This allows the service authority to set the local rules and conditions through bye-laws and contract clauses, within the context of national frameworks. In theory, where the service provider fails to meet the contractual stipulations, it can be held to account by the authority. This type of arrangement is found in **Uganda** and **Burkina Faso**. In the later case each *Commune* can put in place contracts with its operator(s). In **South Africa**, each WSA should have a service contract with its WSPs. The WSA is regulated and monitored by the national DWA (see Figure 7, p84). Whereas this is a simple model, there are limitations to its applicability particularly with CBM approaches when there may be questions over the legality of the water committee as a recognised operator. For example, the system relies on direct communication between customers and local authorities, which hold the contract so that the former can escalate claims and complaints that cannot be resolved locally; however this is not always the case. In South Africa a WSP may have local operators in each village where it provides services, but there may be delays between the operator and WSP in case of break downs. New and wide-spread mobile phone technology is quickly filling this communication gap.

A third way, and a conceptually different modality, is the establishment of independent regulators. In such a case, the regulator sets the 'rules of the game', but relies on other local entities to carry out monitoring and enforcement tasks over the service providers, including the possible involvement of local governments as the service authorities. Three of the studied countries have such independent regulators: **Colombia**, **Honduras** and **Mozambique**, along with the Maharashtra state in **India**. The establishment of independent regulators finds its roots in regulation for privatised urban service providers, particularly in Latin America. And, in fact, many of the initial efforts by these regulators have gone into developing regulation and enforcement mechanisms for urban operators. Regulation for rural areas is only now coming into the picture. **Colombia** is probably the most advanced in this. It has established both the CRA as a regulatory body and the SSPD as a control body. The former sets regulations (for example on tariffs), and the latter carries out the control and enforcement of these regulations among operators. This has proved to be a challenge for rural areas. First of all, there is the sheer number of 12,000 CBM service providers, all of which, in theory, would have to be regulated. Additionally, the type of regulation is not considered appropriate for rural operators by many sector players. For example, CBM operators would need to comply with all the same detailed regulations as large urban utilities, many of which are not relevant. Furthermore, reporting requirements to the SSPD are too onerous for rural operators. Last but not least, in **Colombia** the approach has been considered too punitive, as operators can receive heavy fines in cases of non-compliance. Out of fear of fines, many rural operators have opted not to establish themselves legally, thus avoiding the regulator. AQUACOL and others have engaged in a dialogue with the CRA and SSPD to develop regulation that is more appropriate for rural operators.

In **Honduras** ERSAPS is the regulatory body. It has also focused on urban areas and intermediate towns. It is only recently focusing on regulation for rural areas, and tries to follow a different approach in these areas. ERSAPS aims to transfer the oversight responsibility to a local body, called the Local Control and Supervision Unit (USCL), composed of local government officials and civil society representatives. These bodies would, probably supported by a technician, carry out proactive control over rural operators, which would have to comply with a more limited set of operational indicators than urban operators. The USCLs would also work closely together with the TOMs to provide post-construction support. However, the establishment of these is still in progress, and it is too early to assess them. Plans for regulation for rural areas in **Mozambique** are even more incipient.

In some countries we find partial regulation, with only some parts of the service delivery being regulated, or regulation is spread out over different agencies and levels of governance. For example, it is common that the Ministry of Health is the apex body for issues relating to water quality and regulation of supplies. In **Thailand** this is the case, where the Ministry of Health regulates water quality, both by setting standards and carrying out spot checks. In the **USA** a complex system is established in which the EPA at national level has ultimate responsibility and defines overall regulation, in consultation with State agencies. The local government at county level is responsible for enforcing water quality regulation, but also monitors capacity of operators (for example through operator certification). The enforcement of regulation over tariff setting is split between Water Boards, which are responsible for regulating public and community-managed utilities; and the Public State Utility Commission, which regulates private operators. The disadvantage of such partial regulation is that it is difficult for the consumers to know where to report problems.

The studies highlight an emerging distinction between regulation of the service provider and regulation of the service provided. This distinction is relatively new for the rural sector, although it is a commonly accepted part of urban regulation.

- Regulation of the *service provider*. This refers to the monitoring of aspects such as technical and administrative capacity, financial management and post-construction support. Such regulation is found in the **USA**. **Honduras** and **Colombia** have elements of regulation of the provider in their approaches. Such systems allow for 'positive' regulation, i.e. encouraging improvement of the capacity and performance of the service provider in cases where they may score poorly on certain criteria. This is very strong in some states of the **USA** where there is a clear link between regulation, monitoring and post-construction support, so as to gradually improve service provider capacity. This is different from a more punitive or negative approach to regulation, where CBM operators might be fined when they do not meet certain capacity requirements. Such a punitive approach deters many CBM operators from working

inside formalised frameworks out of fear of overly onerous regulation. As a result they cannot access programmes of support and cheaper capital, and they often remain in a cycle of no regulation and continued poor performance.
- Regulation of the **service provided**. This refers to setting service provision standards in terms of, for example, service levels (including water quality), tariffs and rates, so that the quality of the service that consumers receive is what is regulated – rather than the provider.

As with regulation of the urban sector, and as rural operators are increasingly professionalised, it will be important to regulate both dimensions of service delivery. This implies not only having the regulatory frameworks in place nationally, but the capacity to provide oversight and enforcement, whichever institutional actor is responsible. The cases show that these are still nascent, even in more mature rural sectors as in **Honduras**, **Colombia** or **South Africa**. However, some key lessons are already clear:
- There is need for positive regulation. Holding service providers, particularly rural community-based operators, to account should not only mean having to punish poor performance. This may, in fact, be counter-productive. Rather, it should be about strengthening capacity of operators in those areas where performance is poor.
- This calls for a strong link between post-construction support and monitoring activities. Interventions to strengthen direct accountability can be relatively easily included in post-construction support activities.
- As in the urban sector, regulation should ideally focus on the service provider as well as on the service provided. The more comprehensive this is, the more problems will be detected and/or anticipated. However, care should also be taken not to over-regulate, and to gradually build 'light' regulation that is appropriate for rural operators.
- There is no level in the institutional framework where regulatory functions are best placed. However, the monitoring and enforcement roles need to be present at the decentralised level because of the large numbers of providers involved in rural water supply and the physical challenges this brings. This can be achieved by deconcentrating or delegating certain functions to local government, or by making local government directly responsible for regulation and enforcement.

Monitoring and information management

Monitoring and information management is defined as the collection of data for improvement of planning, implementation and reporting. But, as will have become clear from the previous section, monitoring is also closely related to regulation in that it can supply the various actors involved in regulation with the relevant information to determine whether or not a proper service is being provided at an appropriate price. Whereas regulation is about setting the rules for service provision and enforcing them, monitoring is the instrument that makes it possible. Historically many of the monitoring activities were carried out under development partner funded programmes, whereby individual projects implemented as part of a broader package, or area-based programme, would be visited over time. With the growing role for government, particularly at the decentralised level, many countries have now established national monitoring systems, which in theory collect information on the performance of systems on a regular basis. In practice, collection of such data is often done by a combination of local government and (deconcentrated) ministry staff, and is not always systematic.

In terms of the monitoring function, the cases can be split broadly into two groups of countries: those where monitoring largely focuses on implementation of systems, and those where it focuses on the service provided.

In most of the least developed countries, where the sector focuses on implementation, it is no surprise that monitoring is also focused on implementation or inputs and outputs (e.g. number of systems constructed and number of people served), rather than on the delivery of services or outcomes (e.g. reliability and continuity of water services). Where increasing coverage in hardware is the key sector driver, monitoring is focused on that and not on the performance or sustainability of services, as for example in **Mozambique** and **Ethiopia**. But it is interesting to note that in both these cases, there is a new drive to improve the scope of monitoring, especially in **Mozambique** where the DNA is working closely with the National Statistical Institute, with support from UNICEF, to pilot and roll out a more comprehensive system. Monitoring activities would include aspects such as status of water points (functionality), frequency of maintenance, financial contributions and community participation in O&M. This task may lie with local government, so that during construction corrective actions can be taken. However, data also flows to national agencies.

Countries where monitoring is more comprehensive and is about monitoring actual service provision include, for example, **Honduras** and **Colombia**. Notably, in both cases, the responsibility for monitoring doesn't necessarily lie with local government. In **Colombia**, service providers report directly to the SSPD, and in **Honduras** the task is carried out by the deconcentrated offices of SANAA. As part of the **Jalanidhi** model in the **Indian** state of Kerala, a system for evaluating sustainability was incorporated as part of monitoring efforts. The Sustainability Evaluation Exercise was carried out, but only as a once-off exercise to look at water source, institutional, technical and financial parameters using participatory methods (World Bank, 2009).

Use of sector information management to improve performance

Collection of data on system performance and other criteria in and of itself is only one step in the process of improving the overall situation. The key benefits come through subsequent steps which move from measurement, or collection of data, to the analysis of such data and, finally, to performance management involving the improvement of sector practices, policies and resource allocation (Thompson, Okuni and Sansom, 2005). Put simply, the effort of collecting data through carefully selected indicators is only worthwhile if remedial action can be taken to decide how and where to change practices, invest resources, provide back-stopping support or look into particular issues or trends in more detail. This can happen both at national sector level or at lower decentralised levels such as provinces, departments or districts. Only a small number of countries have incorporated such performance management into the fabric of their national WASH sectors.

Monitoring information from service providers at local level is typically channelled upwards and compiled into national level statistics to be used for macro-level planning. Information management systems have been set up in many cases; for example the District Monitoring and Evaluation (DiMES) system in **Ghana** or the Integrated Management Information System (IMIS) of the DDWS, Rajiv Ghandi National Drinking Water Mission in **India**. The development of such systems is also linked to the trend towards joint sector

Table 9: Uganda's eleven golden indicators

Indicator
1. **Access:** % of people within 1.5 km (rural) and 0.2 km (urban) of an improved water source.
2. **Functionality:** % of improved water sources that are functional at time of spot-check (rural). Ratio of the actual hours of water supply to the required hours of supply.
3. **Per capita investment cost:** Average cost per beneficiary of new water and sanitation schemes (US$).
4. **Sanitation:** % of people with access to improved sanitation.
5. **Water quality:** % of water samples taken at the point of water collection and waste discharge point that comply with national standards.
6. **Quantity of water:** % increase in cumulative storage capacity of water for production.
7. **Equity:** Mean Sub-county deviation from the district average in persons per improved water point.
8. **Handwashing:** % of people with access to (and using) hand-washing facilities.
9. **Management:** % of water points with actively functioning Water and Sanitation Committees/Boards.
10. **Gender:** % of Water User Committees/Water Boards with women holding key positions.
11. **Water resources management compliance:** % of water abstraction and discharge permits holders complying with permit conditions (note that data currently refers to permit validity only).

Source: MWE/DWD, 2010

review processes and the production of (annual) reports to assess achievement of sector goals, and to prioritise themes and areas of focus.

While almost all countries have some form of monitoring system and database at national level, on paper at least, not all function in practice, and certainly not all are based on reliable information. One interesting example is from **Uganda**, which includes information on functionality, and is based on the so-called eleven 'golden' indicators (MWE/DWD, 2010) (see Table 9). Information for these indicators is collected by the districts. However, each district is free to use the information system as it deems fit. The data is collated, first at regional and then at national level, to arrive at statistics that measure progress against the main indicators. Annual targets are set for each indicator in such a way that constant improvement in performance is built into the monitoring framework. The more comprehensive SIAR is found in **Honduras**. The TOMs input information from visits to the communities into this information system, which enables classification according to a range of scenarios from well-performing ('A' category) to non-functioning and in need of major external support to rehabilitate ('D' category). A dedicated worksheet for each community assists in taking appropriate remedial action for each of them. It also allows the analysis of regional and national statistics to be used in defining priority intervention areas.

Having a theoretical commitment to collecting data about implementation of systems and, in some cases, the delivery of services is one thing; actually collecting it in practice (at the designed frequency) is another; and analysing and using this information for feedback into improvements and better performance is quite yet another. Experience suggests that each step along this chain (commitment to monitor; actual monitoring; use of data) is considerably less likely to happen at each step.

Looking again at the **Uganda** case, by having a national level system that reports on a limited number of key indicators, the sector has been able to set out league tables with performance targets in each area, and then carry out trend analyses over the years to pinpoint key issues or apparent bottlenecks across different geographic areas. This, in turn, has led to the commissioning of more in-depth studies to investigate the causes of good or poor performance. Having this type of comprehensive data available, and the trends and analysis based on historical comparison, has enabled the Ugandan sector authorities to take corrective action and to share information, both nationally and at district level, including the links between monitoring sector expenditure (made through condition grants to decentralised government) and performance. Information on and analysis of the golden indicators is shared on an annual basis through a common Joint Sector Review with a broad group of development partners and other stakeholders[4].

It is also true that there are sometimes different, and conflicting, sources of data as is the case in **Ethiopia**, where presently there are often differences between methodologies and results provided by the *woreda-level* surveys and those provided by the regional or national level. Such disputes in information can take on a significant political dimension, as was the case in a recent 'value for money' audit report carried out by the Office of the Auditor General (2009) in **Uganda**. This report disputed the government's own finding of functionality levels of 55% based on spot check samples of systems, as opposed to the level of 82% as stated in the Sector Performance Report of 2008 (MWE/DWD 2008b). Such differences in interpretation of data can also be explained by the differences in perspectives of the entity responsible for gathering data, and the use of different benchmarks in sampling or monitoring techniques.

Whatever the reliability or accuracy of data collected and published nationally, access to this information, especially at lower levels (i.e. sub-national and decentralised levels) is often problematic. Even in cases with quite well developed systems in place such as **Honduras**, accessibility for non-state actors is quite limited. In **India** the Right to Information Act of 2005 enables access to all information collected by state and federal government – at least in theory. This is one of the most comprehensive systems available, and the database of the Rajiv Ghandi National Drinking Water Mission holds on-line monitoring information regarding coverage, progress against targets and budgets (i.e. real against planned expenditures), as well as policy and research documentation. In practice, not all the information is readily available nor updated[5].

Post-construction support

The importance of post-construction support, particularly for CBM, has been highlighted for some time in the literature (Lockwood, 2002; Moriarty and Schouten, 2003; Harvey and Reed, 2006; Whittington et al, 2009; RWSN, 2009), just as it is increasingly being recognised for community-level water resources management (Rivas Hermann, et al., 2010). The notion that once built, systems can be simply handed over to communities, and that the systems will continue to function more or less indefinitely, is now well and truly de-bunked. It is increasingly recognised that some form of post-construction support is an integral part of any SDM. Harvey and Reed (2006) go as far as stating that without such post-construction support, CBM is not viable and other alternatives, such as self-supply or private sector supply, should be considered.

The studies show that there is the need to distinguish post-construction support provided directly to local service providers (including both CBM entities and the private sector) from the broader *capacity support* for local government staff and institutions dealing with rural water. This latter form of support is qualitatively different; it is focused on the service authorities themselves, and is typically provided by central ministries or deconcentrated agencies of such ministries operating at regional or provincial level. Both elements of support can play a role in improving the capacity to monitor. In some cases information collected as part of the support process can be used for regulatory purposes. Table 10 describes a number of more generic aspects of these two different forms of support across the country studies.

Table 10: Generic characteristics of post-construction and capacity support

	Post-construction support to communities and other service providers	Capacity support arrangements to service authorities
Who provides the support?	• Mainly local government staff from district, *Commune* or municipal authorities • Associations of local government (to achieve economies of scale) • NGOs and charities • Associations or confederations of water committees or water user associations • Central government agencies or parastatal entities	• Normally central ministries or agencies responsible for water, and provided through deconcentrated offices • Parastatal institutions • Private sector companies under contract • Large NGOs and charities • Training and academic institutions
Type of support offered	• Technical back-stopping and advice • Administrative and financial • Audit of accounts • Organisational and conflict resolution • Creating linkages with other state and private sector suppliers • Water quality monitoring • Hygiene promotion • Training and refresher courses • Information collection and collation	• Specialised back-stopping and assistance • Capacity building and training • Quality control and adherence to national norms, standards and guidelines • Planning and management, including financial planning • Information collection and collation for national database

Table 11 shows the different arrangements for both post-construction and capacity support across all 13 countries in the study. The details of capacity support are further discussed in the section 'Learning and sector capacity building', p122.

It also shows that in all but a very few cases, there are formally mandated roles for provision of follow-up to communities. In most cases this role is played by local government, as is the case in **Uganda**, **Ghana**, **South Africa**, **India**, **Sri Lanka**, **Colombia** and **Thailand**, which may contract a dedicated entrepreneur or agency to carry out these tasks. Despite this formal mandate, in most cases post-construction support has generally not been applied systematically as an integral part of CBM, even when it is a clear requirement of sector policy. Largely as a result of poor resourcing, this type of support from formally mandated government bodies has been mostly done on an *ad hoc* basis and only when there is a strong demand from the community or operator.

Table 11: Post-construction and capacity support

Country	Post-construction support arrangements for community and other service providers	Capacity support arrangements to service authorities (local government)
Benin	Communes are responsible for guaranteeing service delivery, but often lack the skills and resources to provide back-up support to either the Water User Association or local private operators.	Deconcentrated offices of the Water Ministry at departmental level are responsible for capacity support. Under decentralisation a special programme was set up to develop local authority capacity in areas such as tendering, contracting, management and improved monitoring in three communes within one Department.
Burkina Faso	Communes are responsible for post-construction support as part of their broader mandate as the service authority. In reality, capacity and resource constraints mean that there is limited direct support for communities and this is done on an *ad hoc*, demand-driven basis.	Regional level deconcentrated offices are supposed to support communes, but until very recently there has been no representation of the Water Department at this level. New initiative to bring in engineers to fill gap at regional level to provide support to communes in a range of issues. In addition, there is an institute dedicated to training of water technicians and professionals. *Centre Régionale pour l'Eau et l'Assainissement à faible coût* (CREPA), NGOs and universities provide *ad hoc* support through coaching and training.
Colombia	There is a unit at national level providing direct support, but with limited reach given the large size of the country. There is no overarching national strategy for support to CBM entities by departmental or local government, but some do provide such support directly (e.g. in Cali) or negotiate with the urban utility to do this on their behalf (e.g. in Manizales and Medellin). Several other examples of support models exist; these are much more demand driven based on associations or groupings of associations, including AQUACOL and the National Coffee Growers' Association.	There is no clearly articulated national strategy for capacity support. *Ad hoc* and *de facto* support is provided at departmental level through some large departmental water supply programmes such as the *Programa de Abastecimiento de Agua Rural* (PAAR) (the Rural Water Supply Programme).

Continued ▶

Continued

Country	Post-construction support arrangements for community and other service providers	Capacity support arrangements to service authorities (local government)
Ethiopia	*Woreda* staff provides direct support to WASH committees, but in practice have very limited resources; support based on *ad hoc* arrangements, mainly on demand basis.	Zonal and regional offices of the Ministry of Water are supposed to provide support to *woreda* staff, but in practice this is also very *ad hoc*, and sometimes support is provided directly to WASH Committees.
Ghana	DWSTs made up of district level representatives of relevant line ministries are supposed to provide post-construction support and carry out district planning. But, in practice, their relations to DAs are ambiguous, and they lack operational financing. In theory, support is provided on a demand basis but, in practice, only if there is an ongoing project in a district; and the means available to the DWSTs are too limited to fulfil their function. Direct post-construction support to rural point sources is provided by pump mechanics who are paid by communities.	The CWSA had 10 regional offices, and staff is mandated to support DWSTs with capacity building and training. In practice while well-resourced in terms of human capacity, the regional CWSA offices only operate effectively when there are projects ongoing in their region to which they provide operational and logistical support. Also universities and various NGOs play a role in supporting districts.
Honduras	One main government mechanism is SANAA, using the circuit rider approach. Several other examples of post-construction support models are more demand driven and based on associations or groupings of associations, including: AHJASA and the AJAMs.	Capacity support to municipalities is largely done on an *ad hoc* or project basis, and not as part of a sector-wide, systematic programme. In addition, municipalities support each other and seek capacity through association in *mancomunidades*.
India	Support to service providers varies from state to state. In Gujarat the WASMO programme provides post-construction support (including monitoring, documenting, capacity building, water quality and O&M incentives) to the *Pani Samitis*. In Tamil Nadu post-construction support is provided by the state level Water Supply and Drainage Board to the *Gram Panchayats*, where the TWAD is approached to report problems.	Capacity building (including exposure visits) is carried out for community groups on a range of issues. The best *Gram Panchayat* in each district are used as Key Resource Centres for other *Gram Panchayats* in the district for skills development and O&M training. There are also block-level 'mother *Gram Panchayats*' that are used to support *Gram Panchayats* in need (one each in 243 sub-district 'blocks').

Continued ▶

Continued

Country	Post-construction support arrangements for community and other service providers	Capacity support arrangements to service authorities (local government)
Mozambique	No formal system or mandate for post-construction support for the DAS team. In theory this system is demand-responsive (i.e. District Water and Sanitation Teams [DWSTs] respond to direct requests for help from communities), but in practice there is very limited capacity for follow-up and certainly no funds available for direct support.	DPOPH is responsible for the capacity support role as well as coordination and supply chains, but has limited capacity. There is a technical training institute providing training to local technicians.
South Africa	In case the services are provided by CBOs; the WSA may contract an SSA responsible for post-construction support to CBOs. However, as CBM is hardly ever chosen as the service provision option anymore, this is not so relevant. For other WSP options there are also direct support options, particularly when the WSP is not performing adequately.	Provincial (deconcentrated) offices of DWA play a technical capacity support role to the WSA, providing a 'one stop shop' covering a range of technical, managerial and administrative issues to support a WSA capacity building plan; it is well structured and systematic, with dedicated funding to support local government.
Sri Lanka	Support to communities provided by the RWSSCs [1] at district level; technical repairs still carried out (on demand and payment basis) by NWSDB. RWSSU at district level provides back-up support to *Pradeshiya Sabhas*.	In NWSDB areas, partner organisations are engaged by the Board to provide capacity building support to community leaders, CBOs and local officials. NWSDB also provides a regulatory function with respect to water quality. In CWSSP areas, the Umbrella Management Unit provides support to district level units.
Thailand	Post-construction support to village water committees provided by local government authority or TAOs; in other cases TAO is direct and/or joint operator of systems and therefore receives higher-level capacity support.	There is capacity support to the TAO as service authority by different agencies at national and regional level, including the DWR (provides training, production and dissemination of training, handbook and guidelines, and technical support through 10 regional offices); the Department of Groundwater Resources (gives technical support) and the Department of Local Administration (DLA) (supports TAOs in budget allocation and overall management).

Continued ▶

Continued

Country	Post-construction support arrangements for community and other service providers	Capacity support arrangements to service authorities (local government)
Uganda	DWO staff mandated to provide back-stopping to communities. In some cases, area-based mechanics are in place to serve a number of sub-districts for regular maintenance and to solve minor technical problems; more complex repairs are supposed to be done by the DWO. In theory DWOs have up to 8% of their budget for support to communities in O&M, but in practice this is not always done in a systematic way with staff responding to demand for repairs on an *ad hoc* basis.	MWE has deconcentrated representation at regional level through TSUs which provide support to district staff. The TSUs have a regular programme of support. But with so many districts (and with new ones being created all the time) this supply-driven approach mainly addresses the most under-performing districts.
USA	Post-construction support and technical assistance is provided by two principal organisations – RCAP and NRWA (a membership-based organisation that organises the circuit rider programme). Both are 'bottom-up' organisations providing support to members, but are equally well-linked into government (funding) systems both at federal and state level. RCAP provides more supply-driven support, whereas NRWA tends to respond more to demand.	Both RCAP and NRWA are supported by federal and state funding and also receive direct and indirect support though a number of academic and research centres. These various centres, agencies and programmes research and develop financial planning and asset management tools and training materials, and provide technical advice and support.

A recent study from **Bolivia**, **Ghana** and **Peru** (Bakalian and Wakeman, 2009) highlights the fact that, in spite of these weak formal mechanisms, most communities were seen to be able to access some form of post-construction support on an informal basis, i.e. through wealthy individuals, local politicians or simply through self-help. In whichever way communities and operators are getting help, this type of demand-driven response by its very nature tends to limit the possibility to anticipate problems at an early stage. It therefore becomes a responsive mechanism to solve problems after the fact. This is, for example, the case in **South Africa** where support to service providers (whether community-based or not) follows more of a problem solving route.

There are few exceptions where more structured programmes for post-construction exist. A notable case includes **Honduras** which has had a long-standing programme of support to communities provided through SANAA, and based on the TOM circuit rider model developed in, and adapted from, the USA. These TOMs regularly visit all water committees in their area, and check on the performance of the water supply systems as a basis for identifying corrective actions. Other countries in Central America have similar programmes, such as **El Salvador** (Kayser, et al., 2010) and previously in **Nicaragua** (Lockwood, 2002). However, this TOM programme now faces an uncertain future with the transition of authority to the municipalities for support to the communities, and away from central government.

The **USA** has two very well-established organisations – RCAP (which grew out of six regional NGOs in the 1960s) and NRWA (a membership organisation providing support to community-run water management). Both are 'bottom-up' organisations providing post-construction support to members, and both are well linked to government (funding) systems at federal and state levels. In **Colombia** a mixed situation exists, with the entrepreneurial culture programme for direct support to communities, surprisingly provided straight from the national ministry. This is understandably not very efficient and effective, given the large size of the country; and this national programme has only reached some 10% of all rural operators in its 10 years of existence. There are some examples of municipalities providing post-construction support directly to rural communities (for example, in Cali). An interesting alternative is the contracting out of this function by the municipality to the urban utility (for example, in Manizales and Medellin). Technical staff from the utility then provides technical support to rural operators in the city surrounds, and are paid by the municipality. Such contracted out post-construction support by an urban utility to rural areas is also being experimented with in **Senegal** under arrangements with the Senegalese National Water Company.

In spite of the few programmatic approaches to post-construction, the evidence of their impact is quite positive. For example, the SANAA programme of circuit riders has been able to elevate the percentage of water systems in **Honduras** classified as A from 7% when it started in 1986 to 41% in 2007. Kayser, et al. (2010) show that rural water supply systems in **El Salvador** that are visited regularly by circuit riders have a higher performance in, for example, financial administration and customer relations, and tend to provide better quality water than those who go without support. Rivas Hermann, et al. (2010) also highlight the importance of 'third parties' from outside the community in mediating in intra-community conflicts, e.g. between domestic and irrigation users.

In the absence of structured government post-construction support and the limited responsive capacity of local governments, there have also been post-construction support mechanisms developed by civil society groups and the private sector. Particularly reflected in experiences from Latin America, is the notion that communities can provide mutual self-support through joining together in horizontal organisations and, in that way, professionalise and increase their capacity. There are well-documented cases from **Honduras**, including the Honduran Association of Water and Sanitation Boards (AHJASA) and the Municipal Associations of Water Boards (AJAMs), which provide support to member organisations. Through membership fees, the AHJASA can contract technicians which operate in a similar fashion as TOMs. In addition, both

AHJASA and the various AJAMs can act as mechanisms for rural operators to speak with a common voice in their interactions with (local) authorities (see also WSP, 2004). AQUACOL in **Colombia** operates on a similar premise, with a difference that their mutual support is on a voluntary basis. AQUACOL also fulfils an important role in advocating for the interests of its members with national agencies such as the regulator. Taken collectively, these experiences above all show that communities do have a demand for technical support, and that in the absence of government, they can organise it to some extent for themselves.

These examples of post-construction support provided to – and in some cases by – communities show the importance of differentiating between cases where this is based on supply-driven approaches (such as with the TOMs in **Honduras** and RCAP in the **USA**) and demand-driven support (such as in **Ghana**, **Burkina Faso** and with the associations of CBOs). Both have advantages and disadvantages. Supply-driven approaches may not be very welcomed by communities, afraid of potential regulation; but the main advantage is when organised and funded properly they can pre-empt poor performing systems and identify specific factors that may lead to a breakdown in service (i.e. medium-term maintenance tasks, low financial reserves, etc.). In this sense they represent the ideal, beyond a simply reactive demand-driven model of support.

Discussion: authority functions to facilitate service delivery

Adopting an SDA means having systems and capacity in place at the decentralised level to support different management options so that they can continue to effectively operate, administer and maintain rural water systems. The results from the country studies illustrate a spectrum of how these functions are understood and provided for in practice. On one end of the spectrum there is a situation where many of these functions are still geared towards the *capital intensive* part of the life-cycle, which largely involves increasing coverage through delivery of new or rehabilitated hardware and training. In this scenario many service authority functions are geared towards immediate outputs: planning processes are primarily concerned with new systems, with little consideration for the full life-cycle needs of the system, including water resource requirements; monitoring is focused on progress in constructing; and post-construction support, if considered at all, consists of a few months' support after completion of the project. Not surprisingly, this mode of water supply services provision is most prominent in those countries where coverage is low, and the whole sector is geared towards increasing coverage mainly through implementation of new systems, for example, in **Ethiopia** or **Mozambique**.

The other end of this spectrum includes examples of these service authority functions with a much greater emphasis on service *delivery*. This is illustrated by planning that covers both the implementation of projects, as well as support to existing systems; monitoring of not only progress in outputs, but of the service provided, and even aspects of the performance of the service provider (and, in some cases, a basic form of regulation of rural services); and a greater element of resources dedicated to post-construction support. This scenario tends to be the case in countries where coverage figures are reasonably high, and where the performance and quality of the service has become a concern, rather than simply coverage rates; examples include **Colombia, USA, South Africa** and **Thailand**.

Of course, these two scenarios represent extremes of the spectrum, and many countries show a mixed picture, with better progress in some areas than in others. In almost all cases, the fulfilment of service authority functions is hampered by a lack of adequate financing and technically qualified staff.

In addition, the results show that although many of the service authority functions are carried out by local government, in some cases higher level entities can and do play an important role; for example, water resource planning that cuts across a number of local government administrative boundaries is often taken up by watershed authorities (this is the case with the relatively newly established Regional Water Administration bodies in **Mozambique**). There are also examples of 're-centralisation' of certain functions through the *manocomunidades* for example, in Latin America, which share the costs of specialised technical inputs, and can reach greater economies of scale than individual local government authorities. Other functions require a hierarchical nesting of functions: village level intervention cycles that are embedded in strategic planning at district level, which in turn forms part of regional or national level macro planning.

The adoption of an SDA requires that functions are clearly defined (whether they are carried out at the local government level or not), and that roles are allocated to different actors and supported by legal frameworks. In order to improve sustainability, service authority functions should ideally move along the spectrum from a focus on construction of systems, towards one of supporting a service. Ideally, these functions are mutually aligned and supportive, for example, in such a way that monitoring activities can help strategic planning or targeting post-construction support. The studies show that local government will, in nearly all cases, fulfil many of these roles, but will themselves require support from authorities at higher levels.

An enabling environment for service authorities

The concept of an 'enabling environment' has been recognised for many years in the WASH sector. It is mainly concerned with a range of national level mechanisms and instruments, including policies, institutional frameworks, funding mechanisms and legislation, all of which are necessary to form the basic building blocks for service authorities and service providers to fulfil their functions. The following section focuses on the processes and structures for policy and strategy development, rather than the content, as well as other functions of the enabling environment, including financing, learning and sector capacity building, and harmonisation and alignment.

Policy and strategy development

Developing clear policies, legal frameworks and strategies is a basic responsibility which lies with national level stakeholders (or state level in the case of federal states). As a more service delivery oriented culture emerges it can be expected that such national level policies would be revised to include elements such as specifying sector targets and indicators, clarification of roles and responsibilities in the institutional framework, description of approaches for service delivery and financial frameworks. New or modified legislation is needed sometimes to support these policy positions as they evolve over time.

The case studies highlight a number of examples – both positive and negative – of the process of policy reform. In view of the challenges faced in decentralisation and resistance by previously centralised agencies, a key role in promoting reforms is played by both national policy makers and, in some cases, donors. One of the most commonly cited success factors in the more positive experiences with decentralisation and sector reform is the presence of a strong national vision or strategy and catalytic (political and bureaucratic) champions. Experiences from Gujarat state in **India**, from **Uganda**, **South Africa** and **Thailand** all show how having such clear leadership can accelerate the process of change and support unambiguous policies. These findings corroborate the conclusions from an earlier study carried out by the World Bank (Davis and Iyer, 2002). The corollary of this is that where there is weak or ineffective government leadership, and an absence of champions for the rural sector, it is likely that reform processes, even with significant funding, may be less effective. The challenge for more effective reform processes is therefore one of stimulating better and more visionary sector leadership.

In some cases external support from donors has helped sector reforms to unfold in a systematic way and has provided catalytic financial support for important pilots and innovation. For example, in **South Africa** and some states in **India**, such as Gujarat where bi-lateral funding from the Netherlands government allowed for the development and expansion of WASMO, which has underpinned a successful wave of decentralisation since 2002 in more than 15,000 out of a total of 18,000 communities in the state). In **Uganda** the government's own vision for sector reforms and decentralisation attracted significant bi-lateral funding from the Swedish, Danish and British governments, and allowed for the establishment of the Joint Sector Review process in the early 2000s. It also laid the foundation for the subsequent development of the SWAp (SIDA, 2009).

But in the absence of a strong national vision and clear strategies for the rural sector, development partner support can be fragmented and lead to greater confusion. In counties as diverse as **Ghana**, **Mozambique** and **Honduras** different aspects of the sector reforms have been pulled one way and then another by well-intentioned donors who have provided funding which follows a particular geographic or thematic focus, but which is characterised by being heavily 'projectised'. For example, in **Mozambique** there are many large investment programmes (funded by INGOs such as the Aga Khan Foundation and CARE, or donors such as the Millennium Challenge Compact) which provide support for increasing rural water infrastructure, but which do not always invest significantly in or impact on institutional capacity requirements at either provincial or district government levels. In the case of **Honduras**, support to municipalities has been addressed in a piecemeal way by a series of donors each supporting their own group of local government areas in the absence of an over-arching government-led strategy.

One key lesson therefore, is that to affect policy reform and introduce the associated 'nuts and bolts' that will translate policy into practice (i.e. legislation and bye-laws, appropriate norms and monitoring frameworks), there must be an empowered national government to lead reform processes, as well as supportive development partners.

Financing sustainable service delivery

To achieve sustainable services, all life-cycle costs, including CapEx, OpEx, CapManEx, CoC, ExpDS and ExpIDS, need to be planned for and adequately funded (see detailed definitions on p26). This section presents the trends in the way these are being funded from the three commonly recognised sources, namely: taxes, transfers and tariffs. The mechanisms for funding are in themselves an indication of sustainability. If certain funding streams are not sustainable, or if responsibilities for financing are not clearly defined, it is not likely they will be covered.

Capital expenditures (CapEx) – hardware and software

In new (or upgraded) water supply systems capital costs are generally financed either through taxes or transfers, i.e. from government's own resources or through transfers from donors (including NGOs). It seems to be almost universal practice now that consumers in rural areas are expected to make small contributions to these costs, either in cash or in kind (labour, provision of materials, etc). This principle is in line with the DRA which suggests that communities express their demand for services at least partially through contributions to the investment costs.

In reality, the community contribution is in many cases either only a notional sum or is sometimes waived in order to speed up implementation processes. In **India**, as in many countries, political pressure and priorities often determine which villages receive new infrastructure. Furthermore, there are cases in **India** where private contractors hi-jack the process by filling in the application forms and putting up the required 10% capital cost contribution as a loss-leader to win the construction contract (James, 2004).

The policies in most of the countries recognise that capital expenditure investments have both a hardware (i.e. infrastructure) and a software (i.e. including the costs of community mobilisation, demand creation and the facilitation of the implementation process) component. In line with that recognition, there may be rules on how both these components are to be budgeted for. For example, in **Uganda**, a maximum of 12% of the conditional grant, which is the funding channelled from central government to local government to spend on water and sanitation, should be dedicated to software activities. In spite of this, there are still questions on an appropriate benchmark level for software investments (as a percentage of total investments, including hardware), so to avoid it becoming a 'slush fund' or under-budgeted, which may lead to a poor quality intervention process.

Operation and minor maintenance expenditures (OpEx)

Almost universally, consumers are expected to cover operations and minor maintenance costs through regular payment of tariffs or *ad hoc* contributions in cash and kind, if and when these costs arise. And, although this tends to be the case in reality, a number of important caveats need to be made. First of all, there is often not a clear differentiation between which costs are understood to be OpEx and which ones are CapManEx, and hence what consumers are expected to pay. For example, when a major breakdown occurs, is the consumer still expected to pay for repairs, or is the government obliged to step in? Some countries, including **Honduras** and **India**, have developed detailed definitions of what constitutes minor and major repairs; whereas others have left these definitions open to interpretation. In absence of clear definitions there may be confusion between consumers, operators and the service authorities about who is supposed to maintain the asset, without anyone picking up the bill.

A second issue is that tariffs are more often than not defined on the basis of what is considered feasible by consumers, rather than on calculations of what is actually needed to cover the full costs of operation and minor maintenance. Only some countries have tariffs systems that allow for cross-subsidies within a community, even in rural areas, in which a better-off family pays relatively more than the poorer ones through formal tariff rules as, for example, in **Colombia**. In other cases, such rules may not be formalised, but communities may take the liberty to establish such rules. There is less evidence of direct cross-subsidies between different communities, which instead tends to happen via general taxes. An exception is **South Africa** which has established the Equitable Share, a fund to support operational expenditure on water supply paid for out of general taxes, so that poorer households do not have to assume such costs.

Capital maintenance expenditure (CapManEx)

It appears that CapManEx is the cost category that is least clearly understood, much less planned for in any systematic way. Funds raised from tariffs are seldom, if ever, expected to cover asset renewal or large-scale rehabilitation costs. As already mentioned above, in some cases a distinction is made between minor and major repairs, with the latter then falling outside of what consumers are required to cover. Where such clear definitions do not exist, communities tend to rely on relatively *ad hoc* arrangements to pay for CapManEx. Even in the **USA**, communities appear to follow an approach of 'pick and mix' from the various sources of soft loans and grants available from federal and state governments. In other countries, communities often wait for a major breakdown to occur, and then fall back on local government, the NGO which implemented the original project, or donors to cover these much larger costs. In many cases, these funds are then not readily available, and the agency or department approached by the community is not able to respond, leading to long breakdown times and failed services. One reason for this might be the fact that CBOs are hardly ever the formal asset owners, and would expect local government, which is the asset owner in many countries, to take care of asset renewal.

Only a few examples were found in the study countries where communities or service providers try to address large CapManEx in a more structured way; one clear example is from **Burkina Faso** through the 'mutualisation'[6] of costs within a broader service area, which includes various communities and water systems (see Box 6 for the case of the *Association pour le Développement des Adductions d'Eau* [Association for Water Supply Development]). Another example is the CDF model in **Ethiopia** which started out as a donor-driven initiative supported by the Government of Finland, but has since been taken to scale in Amhara State. Financing within the CDF is channelled from the Ministry of Finance through local micro-credit institutions which then pass on funds to communities for investment in their water and sanitation infrastructure. The key premise of the CDF is that communities and their respective water and sanitation committees are responsible for management of funds, thereby increasing the level of oversight over construction processes. Since the CDF introduction in two *woredas* in 2002, the average implementation of water points per *woreda* has

---

Box 6: Shared managment based on 'mutualisation' of costs in a broader service area in Burkina Faso

In Burkina Faso there is an experience which combines 41 systems in small towns and rural villages from 10 different communes into an Association of Communes to provide support and pool together resources, based on the concept of 'mutualisation'. This includes different technical systems and different types of communes. It makes use of a revolving fund as systems are at a different stage of investment life-cycles. The objective of bringing these systems together into one service area is to reduce transaction costs or to ensure a private operator's activity is feasible. The main three processes that are integrated and 'mutualised' are: management through a dedicated Management Centre, the maintenance process through a contract with a private operator, and the Service Control by contracting an accounting company for audits.

Source: based on Zoungrana, 2011

increased from an average of 25 per year to an average of 54.6 in 2005 in 10 *woredas*. This is translated into the ability to serve 215,000 new users every year, keeping up with population growth.

In absence of structured planning to cover these costs by communities, (local) governments sometimes set aside part of their budgets for CapManEx. For example, in **Uganda,** up to a maximum of 8% of the conditional grant that local governments have for water supply can be dedicated to CapManEx. However, it is uncommon to find such service authorities that have any type of asset management plan for rural water infrastructure where these types of expenditures are planned and budgeted for.

The complexity of planning and budgeting for CapManEx is not unique to rural water supply in developing countries. In Europe it also took a long time before asset renewal and replacement was addressed in a structured way. In his historical overview of water services development in Europe, Barraqué (2009) argues that only in the 1950s did water supply become 'a mature business', and only then started to face the challenge of ageing infrastructure and asset renewal. Before the 1950s this was largely done with public subsidies from national and, above all, local authorities, as part of efforts to universalise access to services. After the 1950s such subsidies became scarcer, and other ways to account for asset renewal and depreciation were needed. Three main approaches to meeting CapManEx and asset renewal costs were identified:
- Concentration of utilities at supra-local level. This would be done, for example, through the merger of various small, local operators to cover a much larger service area, with infrastructure of different ages and types, which would allow for the type of pooling and achieving of economies of scale as seen in Burkina Faso.
- Cross-subsidies via earmarked funds. These are generally cross subsidies from urban to rural users, often via general taxes. **France** is a case where such urban-rural solidarity has been applied (Pezon, 2009).
- Bundling water with sewerage and other utility services, such as gas or electricity. This allows for economies of scale and professionalisation.

Another key finding to note from this study from Europe is that that charging contributions to capital renewal in Europe didn't start until everyone had a decent quality of service. Even in the **USA** currently only 51% of the costs for CapManEx and improvement are met from consumer tariffs (Pearson, 2007). This highlights some important lessons for many developing countries today: first of all, it not realistic to expect that communities alone can cover CapManEx costs. This is already reflected in reality, where often the bill is passed on to external agencies, either government or otherwise, or is simply not covered at all. This is not to say that users cannot contribute to CapManEx at all, but full payment of these costs is highly unlikely.

Secondly, it shows that in order to plan and budget for CapManEx a certain level of scale is needed. When these types of costs are incurred they tend to be usually 'heavy and lumpy' (Barraqué, 2009), representing a peak in expenditure that many service providers simply cannot afford to meet in one go.

By considering other units of scale, cost-averaging and cross-subsidy mechanisms, better financial planning can be done to manage assets in a more efficient and effective way. The almost total lack of evidence from the case studies shows that at present, meeting CapManEx is not even on the agenda in most countries, meaning that assets will continue to deteriorate.

Expenditure on direct and indirect support costs (ExpDS and ExpIDS)

The final cost category is ExpDS and ExpIDS. ExpIDS costs are in theory covered out of taxes, i.e. the revenue of the state. However, in many aid-dependent contexts these costs are also funded directly or indirectly through donors. For example, in **Ghana**, the CWSA charges a management fee from project funds from some donor projects. In the absence of this support, government funding in poorer countries is typically sufficient only to meet recurrent costs such as salaries; operational costs such as fuel, vehicles, computers etc., are seldom properly financed. For example, in countries like **Ghana** and **Honduras**, mention is made of the donor contribution to national reform processes and the establishment of national institutions. More interesting is the payment of ExpDS. As discussed in the section on post-construction support, on p103, post-construction support is a crucial element contributing to sustainability of rural water supply services. And yet, obviously such post-construction support comes at a cost as well, and adequate funding mechanisms are needed to cover those costs.

The studies show that consumers may contribute to ExpDS in a minority of cases (for example through membership fees of AHJASA in **Honduras** or AQUACOL in **Colombia**, or the payment of a nominal fee for a visit by the circuit rider in the **USA**). But even in those cases, the ExpDS is largely subsidised either through government transfers or donor grants. In most other cases where structured post-construction support mechanisms exist, such as the TSUs in **Uganda** or the TOMs in **Honduras**, these are funded through transfers. Because of the aid-dependency of the sector these cases of ExpDS have, in effect, been paid for by donors anyway, even though central government may have transferred funds internally. In the case of **Honduras**, USAID is withdrawing from the water sector, and the post-construction mechanism has come under pressure as the government does not have the funds to meet the full costs of supporting this programme.

One of the explanations for the lack of structured funding for post-construction costs is the fact that very few figures – even estimates – exist on what it actually costs or what would be an acceptable benchmark for the costs of an adequate level of support. Among the few sources of information that exist is the work by Gibson (2010), who reports amounts of between 16.42 and 41.53 US$ per person per year for ExpDS in two district municipalities in **South Africa**. Also, the study in **Colombia** attempted to gain insights into the costs of the different post-construction activities, and it immediately became clear that the costs of these programmes vary by orders of magnitude, not least because the type of post-construction support provided is always situation specific. Just as the costs of providing a higher level of water service are higher than for a basic level, the costs of a comprehensive post-construction support package are higher than for once-off visits (see Table 12 with data from **Colombia**). Post-construction support cost is not cheap. As Gibson (2010) discusses from a comparative study in **South Africa**, the costs, particularly of travel and professional staff who can provide such support, is significant, and is much more than previously assumed.

These findings show that a gap exists in terms of the clarity of funding of two cost categories, namely large CapManEx and renewal costs and ExpDS, such as those that enable post-construction support for CBM, or small-scale private operators. To a certain extent this is understandable as these are the type of cost categories that conventionally planners and programme designers do not consider since they go beyond the timeframe and geographical reaches

Table 12: Comparison of costs of post-construction support in Colombia

Post-construction support experience	Type of support provided	Costs
Programa cultura empresarial	Comprehensive package of support by national MAVDT to modernise and professionalise service provider	On average US$17,500 per service provider as once-off investment
AQUACOL	Horizontal learning and exchange between community-based service providers	Membership fee of between US$108 and US$324 per service provider per year to cover technical assistance costs
Comité de Cafeteros	Regular support visits by National Coffee Growers Association to systems of its members, and support in O&M, billing and tariff collection	US$836 per service provider per year
Programa de Abastecimiento de Agua Rural	Social mobilisation, establishment and training of the service provider as part of construction, rehabilitation and extension of systems	Between US$14,000 and US$28,000 per service provider as once-off investment
Agua de Manizales	Urban utility providing capacity training and technical assistance to surrounding rural schemes	Not quantified by the urban utility as it 'merely' concerns utility staff time
Aguas Manantiales de Pácora	Technical assistance by urban utility to surrounding rural schemes, on request basis	Not quantified as it is a 'voluntary' service by the utility to the surrounding villages

Source: Rojas, et al., 2011

FINDINGS FROM THE COUNTRY STUDIES   119

of a 'project'. This lies at the heart of the failings of a 'project-based' approach, and this lack of attention to CapManEx, ExpDS and ExpIDS are only recently entering the debate on the sustainability of rural water supply (see, for example, Fonseca, 2011a).

In this respect, much can be learnt from the urban water utility sector. Ofwat defines 'capital maintenance' as 'how companies are required to maintain the operating capability of their asset systems to ensure continuity of service for current and future customers' (Ofwat, 2005). Most utilities have some kind of asset management plan, in which account is made for replacement of different assets (equipment, installations, etc.) with different depreciation times. Such asset management plans are not common in rural water supply at the level of the service provider, or at the level of the service authority (which in many cases is the actual asset holder), or even at national level. Moving towards more sustainable service delivery requires similar kinds of plans which are also adequately budgeted for, at different levels, beyond the project cycle only.

The covering of these kinds of costs cannot easily be done under a project or programme approach, but requires sector funding, where finances from different streams can be pooled. In some of the countries that are highly aid-dependent, in fact, there is a trend towards sector budgets[7], and sector budget support by donors, as part of a SWAp, including in **Benin**, **Uganda** and **South Africa**. **Burkina Faso**, **Ethiopia**, **Ghana** and **Mozambique** are all moving towards a SWAp as well. This is of crucial importance to start addressing large CapManEx and renewal costs and ExpDS. Further details of these SWAps are provided in the section 'Aid effectiveness', p128.

Size of funding flows

In addition to the types of cost categories that are being met (or not), it is also important to understand whether the size of the funding flows is adequate to cover these costs. In spite of attempts to do so, the study failed to obtain a comprehensive overview and comparison of the size of financial flows in rural water supply in the 13 countries. The reasons for the lack of comparative data on sector investments are manifold:
- First of all, there is no uniformity in the definition and use of cost categories in sector budgets of expenditure overviews. **South Africa**, for example, produces an overview of sector investments, clearly differentiating between CapEx investments and operational costs (which lumps together OpEx and CapManEx), and hence also adds investments by users through tariffs to the sector balance sheets. Most other countries limit overviews of sector investments to capital investment data only, such as for example in **Benin**.
- Secondly, the definition of what is included in the 'water sector' differs from case to case: some countries disaggregate information between the

rural and urban sub-sectors, or between water supply and sanitation, whereas others do not. That makes comparison from one country to another difficult.

- Thirdly, there is still a general lack of benchmarks for unit costs of water supply. As discussed in more detail in Fonseca (2011b), unit costs have only recently come on the agenda of the water sector as part of studies to assess whether current funding flows are adequate to meet the MDGs. As a result, recent studies have started developing this data, but most of it only refers to CapEx and more minor O&M costs. Without insights into unit costs of services in the different stages of the life-cycle these cannot be aggregated at sector level, and there is still insufficient data to understand the adequacy of sector finance towards sustainable water services. Of the study countries, to our knowledge, only **Ghana** and **Uganda** have completed studies into unit costs (KA Associates, 1999;

Box 7: Tracking sector investments in the joint performance report in Uganda

Every year, the Ministry of Water and Environment produces a joint sector performance report for the water sector in Uganda. One of the chapters is dedicated to sector investments. As many of the major donors in the country put their sector funds through a SWAp, alongside government funds, it is relatively simple to obtain an insight into how much funds come from these two sources of funding.

The report from 2010 indicates that, of the on-budget resources allocated to the sector in the Financial Year (FY) 2009/10, the equivalent of US$52.8 million (53.1%) came from the Government of Uganda – a figure which includes grants and loans from development partners that have embraced and are operating under the preferred basket funding mechanism – while US$47.1 million (46.9%) came from development partners outside the basket funding arrangement.

Out of this, US$30.8 million was destined for Rural Water Supply and Sanitation. Some 76% of this budget was allocated to the District Water and Sanitation Development Conditional Grant (DWSDCG). The DWSCG Guidelines for 2009/10 required districts to allocate the grant as follows:

- Office operations – guide 5% (actual 10%)
- Software activities – guide 11% (actual 9%)
- Water supply – guide 70% (actual 72%)
- Sanitation (hardware) – guide 6% (actual 2%)
- Rehabilitation – guide 8% (actual 6%)

UWASNET tries to compile similar information from the NGOs who are members of UWASNET. Despite some difficulties in this, e.g. on definitions on which cost categories are included or not, this at least gives an indicative figure. Private and household investments in water supply are not tracked in this report.

Source: MWE/DWD, 2010

MWE/DWD, 2008a), albeit that these focused on CapEX and more minor O&M costs (CapEx and OpEx) only. The current WASHCost project (Moriarty et al., 2010b) is looking into the other types of life-cycle costs.
- Last but not least, in most countries there are many actors investing in the sector: national and decentralised authorities, donors, NGOs, utilities, private sector and consumers themselves. Tracking and compiling such information on a continuous basis in a comprehensive format is simply not an easy task, even though some countries, such as **Uganda**, are putting in effort to do so (see Box 7). In the **USA** there is a myriad of sources of funding for both CapEX and large-scale CapManEx, but there is no single overview that brings all of these together. At the level of donors, according to the WHO (2010a), it is relatively easy to track government and Organisation for Economic Co-operation and Development (OECD) donor investment, but the amounts being invested by non-OECD donors, the private sector or NGOs, and the amount spent directly by households are less well known and, for these reasons, are typically excluded from country investment overviews.

In spite of the complexity of getting accurate data, a first, and probably not surprising, finding is that the least developed countries are highly aid dependent. Table 13 provides an overview for those countries, for which some data is published, of donor aid (grants) as percentage of expenditure on water and sanitation. All caveats mentioned above apply, such as the definition of what is included in these figures, so the figures should not be used for comparison across countries. Rather, they illustrate the extreme levels of aid dependency in the WASH sector in the cohort of least developed countries. Under such conditions it is extremely important to achieve some form of harmonisation of donor efforts.

Table 13: Donor dependency in the water and sanitation sector in selected study countries

Country	Donor aid as % of sector expenditure on water and sanitation
Burkina Faso	89[A]
Mozambique	87[B]
Benin	76[C]
Ethiopia	49[D]
Honduras	21[E]

Sources: [A] Zoungrana, 2011
[B] WHO, 2010a
[C] MEE, 2010
[D] Chaka, et al., 2011
[E] Serrano, 2007

Secondly, estimates indicate that current investment levels are far below what is needed to achieve targets for both CapEx and CapManEx requirements for rural water supply. Examples are numerous. In **Honduras**, a study on financial flows in the sector revealed that the indicative funding flows are orders of magnitude lower than what would be needed to achieve the MDGs. The same is echoed in the GLAAS report (WHO, 2010a) where 35 of 37 countries report that financial flows are insufficient to achieve the MDG target for sanitation and water.

In order to plan for future investments there is need for having an overview, even if it were approximate, of all sector expenditures and their break down according to cost categories, especially to order to avoid major maintenance and replacement backlogs. In **South Africa** funding has been made available to address the backlogs in capital investments, and coverage figures for rural water supply have increased dramatically. However, there is an increasing need for adequate financing of the maintenance backlog. According to projections by Hutton and Bartram (2008), maintenance and replacement (i.e. OpEx and CapManEx) of existing water supply and sanitation infrastructure would make up 74% of all financial needs to reach the MDGs. Yet, many national governments and development partners continue to favour investing in new infrastructure only. Data from eight major donor agencies shows that 64% of their aid to drinking water and sanitation is disbursed for new services, and only 13% to maintaining or replacing existing services (WHO, 2010a). Even in the **USA** it is estimated that 1.3 trillion US$ is needed over 20 years to address capital maintenance backlogs, an amount that is well above the estimates of what is currently available. In Europe there are also increasing concerns about the escalating costs of asset renewal, and the way to provide for these (Barraqué, 2009).

Learning and sector capacity building

A second function of the enabling environment includes learning and capacity building, which can further be broken down into capacity support to decentralised levels and, secondly, support to establishment of sector learning mechanisms at the national level.

Capacity support to decentralised level

As discussed in service authority functions (page 110), capacity support is understood as the support to service authorities, mostly provided by agencies from the national level, or through deconcentrated offices at provincial or regional level. A good example of these are the TSUs of the MWE in **Uganda** (see Box 8). A similar type of support mechanism is just being set up in **Burkina Faso** through a new programme to establish regional centres to support communes. As well as in **Uganda**, there are relatively well-established capacity support programmes in **Ghana**, **Benin** and **Sri Lanka**.

> ■ ■ ■ **Box 8:** Uganda's technical support units (TSUs)
>
> As part of government's responsibility and commitment towards deepening decentralisation, the Directorate of Water Development (DWD) established regional TSUs in 2002 to build capacity and to provide backstopping support to District Local Governments to be able fulfil their new roles and responsibilities in the provision and management of sustainable water supply and sanitation. The TSUs were set up as a transitional arrangement to raise capacity of districts to manage the conditional grants under the District Water Supply and Sanitation Programme (institutional development) as well as build capacity of service providers for improved service delivery (skills development).
>
> TSUs therefore provide support to districts on a demand-driven basis. The TSUs are a temporary support measure and will be phased out as the local government capacity increases. The origin of TSUs in Uganda can be traced from the broader changes that have taken place in the sector and include decentralisation of rural water supply to districts, consequently changing the role of the centre from direct implementation to policy development, providing support, monitoring, and regulation.
>
> There are eight TSUs (based on the same number of water catchment zones in Uganda), each with staff that includes a Water and Sanitation Specialist, a Community Development Specialist and a Public Health Specialist. Each of these TSUs is headed by a Focal Point Officer or Coordinator. The main areas of support provided by the TSU staff include four core components:
>
> 1. *Planning and management*, including implementation of national policies and strategies, development of plans and development and use of a Management Information System.
> 2. *Quality assurance*, including compliance with national policies and guidelines, management of the tendering and procurement process and supervision of private sector and NGO/CBO to ensure value for money.
> 3. *Capacity building and inter-district cooperation*, including conducting self-assessment and identification of capacity gaps, development of capacity-building strategies, implementation of training activities, promotion of NGOs/CBOs and private sector participation, and promotion of sector coordination and inter-district learning.
> 4. *Specialised technical assistance*, including promotion of appropriate technologies, gender mainstreaming and facilitation of capacity building workshops.
>
> Methods used in providing capacity building include training in classroom/workshop setting, consultative meetings, on-the-job training and demonstrations, provision and interpretation of guidelines, development and provision of formats, quality assurance of plans, budgets and reports.
>
> **Source:** Nimanya et al., 2011

In **South Africa** the deconcentrated offices of DWA have set up so-called 'one stop shops' to ensure access to specialist expertise to assist the WSA in meeting key performance targets. There is a structured cycle to plan for support interventions, as shown in Figure 11.

124 SUPPORTING RURAL WATER SUPPLY

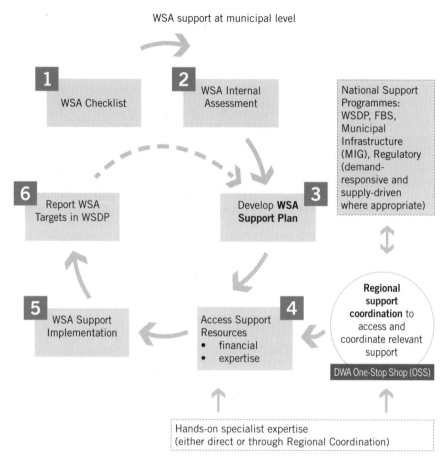

Source: de la Harpe, 2011b

Figure 11: Process for support to WSAs in South Africa

In **Honduras**, in spite of having some very well developed models of post-construction support to communities, the support offered to municipalities is fragmented, and happens on the basis of project funding. In practice this means that some municipalities are supported by one project, others by a second, and yet many more do not receive any capacity support at all. Just as communities come together to share capacity, in **Honduras** municipalities also sometimes join forces through *mancomunidades*. This approach, which is also followed in other countries in the region (such as **Bolivia**), allows the *mancomunidades* to contract specific expertise such as engineering which they otherwise could not afford alone. In addition, they carry out joint planning and implementation by pooling funds.

Whereas the cases above refer more to on-the-job support, some countries also have more formal learning and training programmes. These are provided by a range of institutions and training centres, linked to either government structures, universities or civil society resource centres. These include, for example, CREPA in **Burkina Faso**; WES-Net (Water and Environmental Sanitation Network), a network of resource centres as platforms for learning and new knowledge for state level officials in **India**); and CINARA in **Colombia**, a research and development institute based at the University of Valle.

Despite these examples of capacity support – either through on-the-job training or through resources centres – in most of the country cases studies there are still deep and chronic problems with capacity at decentralised level. Much of this is not even specific to the water supply sector; put simply, there are chronic capacity problems across all areas of local government in many countries. Many local governments struggle to even carry out basic governance functions (tendering, planning, procurement, monitoring, etc.). This is one of the reasons for the low levels of trust or confidence in local government, which may stall decentralisation efforts even more. Hence, the growing recognition of need to link capacity building efforts in the water sector to broader public sector capacity development. The distinction between specific capacity building for the (rural) water sector and broader efforts to improve capacity is highlighted in **Ethiopia** where there is a Public Sector Capacity Building Programme managed by the Capacity Building Bureau of central government, but it is weak on water issues with few links to the sector.

Learning mechanisms at national level

In most of the study countries there are established learning and coordination platforms at national level, both between donors and between government, donors and NGOs, often with working sub-groups addressing specific sector issues. Although such mechanisms have emerged in different ways and with different drivers, two main categories can be identified.

Firstly, there are those mechanisms that are led, or dominated by *government* and form part of the formal sector learning. For example in **Uganda** there is the Water and Sanitation Sub-Sector Working Group, which itself has two permanent so-called 'sub-sub-sector' working groups: one for Water for Production and one for Sanitation. A number of further temporary sub-groups have been established that are mandated to investigate and promote innovation in key thematic areas such as financing, sanitation and technology. In **Mozambique** the water and sanitation group *(Grupo de Água e Saneamento)* of the DNA is a good platform for sharing and coordination, and this model is starting to be pushed out to provincial level, bringing together government, NGOs, development partners and academia.

In part linked to the emergence and growing capacity of such platforms is the trend to develop joint sector reviews. These have been adopted in a number of countries, with **Uganda** being one of the earliest cases, and seek to define progress against a limited set of commonly agreed upon sector performance indicators, which are recognised by government and development partners. This process allows for greater efficiencies in reporting, lessens the burden on government, and provides a greater focus on a particular aspect of performance; for example in **Uganda**, a sub-group was set up to look into the functionality of systems after this was flagged as a critical issue in the 2008 Joint Sector Review. It is interesting to note that such joint sector review processes have emerged mainly in the more aid dependent countries, including **Ethiopia**, **Mozambique** and **Ghana**, amongst others, driven by the need for better donor coordination and harmonisation.

The second broad group of learning mechanisms are those that have emerged from, and are driven largely by, civil society. However, given their related mandates for advocacy, these are not always mutually exclusive from government bodies. Examples include UWASNET in **Uganda**, which provides both coordination and learning functions to its member organisations, composed of a range of NGOs from very small local organisations to the largest international NGOs. In **Ethiopia** the Water and Sanitation Forum is a national platform engaging civil society. It is hosted by the consortium Christian Relief and Development Association, and is involved in capacity building, policy advocacy, and sharing information with its members, including dissemination and support to the new One WASH monitoring and evaluation system. Apart from government, in **India** there is WESNET India and Solution Exchange, both of which are Delhi-based and offer fairly effective platforms for sharing of experiences and information flows. A third entity is India Water Portal based in

---

Box 9: The Honduran network for water and sanitation, RASHON

Red de Agua y Saneamiento de Honduras (RASHON) is a national network consisting of government, NGOs and donor members working on issues of water supply and sanitation in Honduras. Formed in 1990 in part as a reaction to the real need for coordination of water and sanitation initiatives, RASHON has three core strategies: 1) the promotion of information management and dissemination of good practices; 2) advocacy for improved investments in the sector; and 3) institutional strengthening of stakeholders at all levels.

Various thematic working groups have been established as part of its knowledge management initiative. One of these is dedicated to the theme of sustainability. It has evolved over the years to become a respected network in the water and sanitation sector in Honduras and through this role has contributed to the reform and modernisation process in recent years.

RASHON benefits by not only counting NGOs in its membership – both small and large – but by the active participation of the main government ministry (SANAA), the sector policy agency (CONASA) and the national social investment fund (FHIS), all of which help to make it a very broad platform for the sector and increase legitimacy.

Source: López, 2011

Bangalore. Some of the platforms have actually tried to close the gap between government and civil society such as RASHON in Honduras (see Box 9).

Technology development and management

A specific area of relevance for learning relates to the management of technology options that are available in a country. This includes technological innovation, often captured under various names such as appropriate technology and low-cost technology, around which there is a wide body of literature. But it also entails the capacity of a sector to analyse which technologies are considered more or less sustainable in their context, how innovations which may improve sustainability are taken up, and how to manage supply chains for these technologies.

In general terms across the case studies it is evident that conventional communal technologies (such as boreholes with handpumps or small-piped systems) are generally well known and standardised. Examples are the cases of both **India** and **Thailand**, where technical guidelines and costings are provided for a range of different options and sizes of systems. These are used to improve transparency in tendering and construction processes. Also many countries have gone ahead in standardising handpumps, such as **Mozambique**.

This is particularly important in countries where handpumps are the predominant technology, as supply chains for spare parts for handpumps are notoriously fraught with problems. The reports from **Mozambique**, **Ethiopia**, **Burkina Faso**, **Benin**, **Ghana** and **Uganda** all highlighted the difficulties in accessibility of spare parts and setting up sustainable supply chains. A range of supply chain modalities have been tried in these countries – these include making stocks available at regional or district government offices, promoting local private suppliers to equip water committees with a start-up stock, and PPPs. They have all had mixed levels of success. Reasons for failure included the fact that in many countries markets for spare parts were often too small and profits were too small for the local private sector to step in, while the public sector based initiatives often had too limited resources, were too spread out and were not sustainable for the government agencies involved. Oyo (2006), in his review of lessons on supply chains for spare parts for handpumps from 25 studies in 15 countries, concludes that there is not one single approach to spare part supply chain management that works best. Depending on the context, either public, private or mixed initiatives work best. And, as reported in our cases, even then it may work sub-optimally. What is important is that supply-chain management should exist at sector level, which entails, for example, standardisation so as to reduce the fragmentation of the market and making spare parts supply economically viable for suppliers, and also managing processes of piloting different supply chain management models. This is also echoed and elaborated upon in more detail in Harvey and Reed (2004).

Supply chains are reported in our cases as being less of an issue in countries where piped water systems are the main technology option. In **Sri Lanka**, for example, it is reported that most spare parts for piped systems are readily available in local hardware stores. In **Honduras** the only problem experienced is in chlorine used in surface water-fed piped systems. It is often expensive for individual CBOs to procure and store chlorine. Associations of community-based providers sometimes step in by establishing a regional chlorine bank where members can buy chlorine for reduced prices due to bulk procurement.

The cases also show that there might be a flip side to standardisation, as it can become a limitation to innovation. Countries such as **Mozambique** had, until very recently, extremely stringent rules on handpump design. This limited the introduction and innovation of new handpump options. The earlier mentioned sector platforms can provide the space within which to introduce and trial new options, and in which other sector players can carry out reviews of innovations. A similar experience exists with the WASH Cluster in **Zimbabwe** (Makoni, et al., 2007).

Aid effectiveness: harmonisation and alignment

A further trend affecting rural water supply has been the drive towards addressing aid effectiveness. As shown in Table 13 (p121), in some of the least developed countries aid represents a very high percentage of total WASH sector investments. Even in some lower middle income countries, aid still represents a significant percentage of sector investments, as echoed in the GLAAS report (WHO, 2010a). This situation is reinforced for the rural sub-sector, which tends to receive more support from external donor assistance than the urban water sector, which relies more heavily on public investment (Hall and Lobina, 2010). It is therefore not surprising that the rural water supply sector has reflected the main thinking and approaches towards aid delivery over its history.

The predominant modality used by development partners until about the mid 1990s was based largely on project and area-based programme approaches, often working in large geographical areas (such as a province) to increase coverage. In some cases these programmes were able to develop consistent and high-quality intervention methodologies such as, for example, the Rural Water and Sanitation (RUWASA) programme in **Uganda**. Although these approaches allowed for a stronger role for local government, there are many limitations to this approach identified in the aid effectiveness debate. Programmes often had stand-alone implementation units independent from government and its systems, which did not contribute to the building of government capacity for programme implementation.

From the early 2000s the debate on aid effectiveness was revived by continued concerns on issues such as high transaction costs, fragmentation and concentration of aid effort between 'donor darlings' and 'donor orphans', failure to use national systems for managing aid, and the limitations of the aid architecture, where aid conditionality was seen to reduce country ownership and effective results (de la Harpe, 2011a).

These concerns have culminated in a series of political agreements on the aid architecture in an attempt to adopt common frameworks and indicators to improve effectiveness, most notably the Paris Agreement of 2005 and the Accra Agenda for Action of 2008 (OECD, 2008a).

The experiences from the case study countries show – particularly for Africa, where the most aid dependent countries are located – the manifestations of the Paris Declaration and Accra Agenda through different mechanisms and approaches towards improved aid effectiveness in the WASH sector. These include mechanisms such as the establishment of SWAps, alignment by pooling of funding and in some cases direct budgetary support to regular ministry accounts. **Uganda**, **Benin** and **South Africa** are the only three countries in the case studies which have formally adopted a SWAp mechanism to date.

Box 10: Interview results from Ethiopia on the National One WASH Programme

Advantages of taking a harmonised approach:
- Reducing reporting requirements, important in a context of limited capacity
- More effective monitoring using common indicators and a consistent baseline, enabling better measurement of progress
- Funds will be utilised more effectively and unit costs of providing the services will be more efficient
- Standardisation will simplify implementation
- Pooling funds will streamline decision making
- Better access to information with one monitoring and evaluation system in place
- Reduced transaction costs
- More clarity on who is doing what and where based around a single plan
- Strengthened role of NGOs where their contribution to the sector is more apparent and creating opportunities to participate in national dialogue and influence policy processes for pro-poor outcomes

Requirements to make the transformation to a harmonised approach:
- Strong leadership
- Positive attitudes
- Commitment by all
- Strengthened capacities at *woreda* and *kebele* levels
- WASH forums for learning and information sharing mechanisms
- Partnership between government, donors, NGOs and private sector
- Strengthened integration and coordination mechanisms
- Baseline information collection and documentation

Source: Chaka, et al., 2011

However, it is interesting to note that some form of SWAp or common framework for development assistance is also being set up in **Ethiopia**, **Mozambique** and **Ghana**.

In **Mozambique** there is a rural SWAp under preparation which will establish a coordinating secretary to set up platforms at national and provincial level; most donors already communicate amongst themselves through a well-established roundtable and troika leadership model. All donors have now signed a code of conduct for operating within SWAp towards financing (except USAID, the Millennium Challenge Corporation and Japan), and there is an in principle agreement to set up a common donor fund. In **Ethiopia** a sector harmonisation programme known as 'One WASH' started in 2006. It has widespread commitment, and is now gathering momentum. The focus of this harmonisation drive is on going to scale, and sets very ambitious coverage targets. There is the danger that this drive for achieving scale is not addressing critical aspects of sustainable service delivery. The challenges faced by this change process require greater scrutiny and constructive discussion within the sector. The viewpoints from a series of interviews with government, development partners and civil society groups are given in Box 10.

The problems of a lack of a SWAp can be seen in **Honduras**. Recent changes under the sector modernisation process have been promising, and include many of the elements and institutions that are required for adopting a sustainable SDA on paper. However, in practice, these elements are not being brought together in a coherent way or operationalised systematically through an overall sector strategy or investment plan. This is compounded with insufficient funding for key government coordination bodies such as CONASA, which is responsible for planning at national level. As a result, many of the reforms are funded by different financing agencies, leading to a patchwork of donor-funding for different parts of the reform, and 'projectised' interventions.

Most of these efforts have been between governments, the major multilateral financing institutions (the World Bank and the regional development banks) and (European) bilateral donors. The majority of INGOs and smaller 'charity' or philanthropic agencies still tend to operate outside of such frameworks, and follow the more uncoordinated project-based way of working that does not allow sustainability to be addressed beyond the limits of the projects. Having said that, there are also examples of change; for example through umbrella organisations for NGOs including UWASNET (**Uganda**) and Coalition of NGOs in Water and Sanitation (CONIWAS) (**Ghana**); or activities of individual NGOs in certain countries (such as Water for People in **Rwanda** and **Honduras**), Plan in **Guatemala** and UNICEF in **Bolivia**, both of which are providing more structured support to local municipalities (Brussee, Marín and Smits, 2010).

Despite some good examples, progress against the various commitments has been mixed. In its 2008 survey of aid practices in 54 countries, OECD found that only 43% of donor supported projects and programmes were using partner country procurement systems (OECD, 2008b). For the water supply and sanitation sector, the data is even worse. A report issued by the EU Water Initiative, Africa Working Group, shows that still more than two-thirds (71%) of all European financing is channelled through projects and programmes, with about a fifth of all aid classified as 'not coordinated' with national government programmes (Fonseca and Diaz, 2008). The OECD survey illustrated that aid partnerships have experienced severe problems. One of these lies with recipient country capacity to absorb and properly implement funds, with some cases of extremely low disbursements. Although available data is a bit outdated, the report from **Benin** mentioned only 45% of budgets were disbursed in 2006 (Adjinacou, 2011), and in **Mozambique** 28% in 2004 (Munguambe and Langa de Jesus, 2011). Such capacity constraints include resources for operational activities, limited technical and managerial skills, and lack of knowledge regarding sector frameworks, roles and functions, and legislation.

Improving aid effectiveness – and putting in place associated mechanisms for greater harmonisation and better alignment – takes some time to yield results. A recent review of the literature indicates that despite broad acceptance for the principles of the Paris Declaration, and all the high level forums and commitments to improve aid effectiveness, the extent to which donors and multilateral bodies adhere to and implement such principles is mixed. There have been improvements in aid effectiveness in the health and basic education sectors where development assistance is considered better managed,[12] but even here practitioners agree that aid effectiveness has not achieved its maximum potential (OECD, 2008b). Although aid effectiveness modalities and approaches are less developed in the water sector than in the education and health sectors, the perception that water may be 'lagging behind' is without evidence (de la Harpe, 2011a). In the WASH sector the new Sanitation and Water for All – Global Framework for Action is just being established and, alongside the Global Sanitation Fund, both seek to address greater coordination and harmonisation of funding at the macro-level.

Because the rural sector has been so reliant on external donor funding in many countries, this has resulted in a multiplicity of actors which have tended to invest directly in their 'own' infrastructure programmes. Nonetheless, this situation is changing in a number of countries, and while few donors have so far moved to direct budgetary support, more and more are at least working together in a more coordinated way with funding, and supporting common priorities defined by over-arching national planning priorities.

Finally, there are other variables which can affect the progress of aid effectiveness in the water sector in profoundly different ways depending on the country context, including socio-economic, political and other national factors. Political commitment, in particular, is a key determinant that can either advance or obstruct aid effectiveness (Danida, 2006). The mere presence of aid effectiveness modalities does not necessarily mean improved delivery and use of aid assistance. Power relations and incentives underpin policy, financial and institutional decisions which impact upon the overall performance of a sector. A stark example of this comes from the **USA**, the most developed country in the study grouping, where funding flows to the rural water sector are extremely fragmented, and work across a multiplicity of federal, regional, thematic and state sources. There was an attempt in Ohio State to set up a 'mini-SWAp', bringing together all funding streams under one umbrella, and effectively creating a centralised help-desk for rural water operators. However, local politics and the fact that the bureaucracy of financing mechanisms (with rewards for continued disbursement of funds) acted as a disincentive to better harmonisation, meant that the initiative was shut down after a few years (Gasteyer, 2011).

Operationalisation of aid effectiveness at the local level

While there have been considerable efforts to improve coordination, harmonisation and alignment at international and national levels, this does not always translate into improved operational practice at sub-national and decentralised levels. A number of recent studies have indicated that this is the next challenge for the aid effectiveness movement – to ensure partnerships 'reach beyond capital cities' (Welle, Nicol and Van Steenbergen, 2008).

For example, in **Uganda**, which is one of the earliest adopters of the SWAp mechanism, coordination of different development partner efforts is not always easy when it comes to local and even international NGOs operating at district level. While there are examples of good coordination mechanisms at district level between local governments, user groups, NGOs and development partners (such as the DWSCC in Uganda), attendance, performance and impact of these committees is extremely patchy, and often depends on the diligence of district level staff. This is perhaps a reflection of the fact that some 47% of assistance for rural water still by-passes the central government's basket funding mechanism. In many countries, despite the fact that local government has the mandate and authority to coordinate the efforts of NGOs and development partners in the sector, it is often merely informed at most, and completely by-passed at worst.

Weak capacity to really operationalise the 'spirit' of the SWAp at local level can also result from partial decentralisation, and there is evidence from a number of countries in the study that coordination functions are being retained at higher levels. For example, in **Ethiopia** coordination and division of labour between donor and NGO programmes is largely done at regional level and is quite effective, but tends to undermine the authority of the *woreda* government. In the case of **Mozambique**, with extremely limited district capacity, most operational coordination takes place at provincial level under the auspices of the DPOPH, which chairs an inter-sectoral body that incorporates agriculture and health, as well as water.

Overall, the evidence from the cases studies underlines the problem with attempts to improve aid effectiveness in rural water, which is that despite agreement 'from the centre' on working more closely and harmonising through

■ ■ ■ **Box 11:** Tamil Nadu – A multiplicity of agencies at the village level

> In many areas in the state, villages have multiple sources of water. Apart from the Tamil Nadu Water and Drainage Board, water is also provided by the Department of Rural Development (DRD), mainly through handpumps and power pumps. The funding for these interventions is in the form of grants from the Rural Development Fund. In addition, water supply programmes are also taken up from the funds devolved under the 11th Finance Commission, Minister of Parliament and Members of the Legislative Assembly's local area development funds. The engineers under the DRD set-up and implement these programmes. Under *Swajaldhara*, rural water schemes are also provided to villages. The lack of coordination between the Engineers of both departments is evident from the fact that more than one type of scheme is often implemented in the same village. The result is water supply schemes in some villages that are far in excess of the requirement.
>
> **Source:** James, 2011a

mechanisms such as a SWAp, on the ground the situation can be quite different. This has as much to do with the organisational culture of development partners (and their need to seek 'identifiable' interventions), as it does with national capacity constraints.

Inter-ministerial coordination

While there has rightly been a focus on development partners to improve their actions and be more in alignment with government led priorities and processes, there is also a case for improving the recipient government's coordination and harmonisation approaches. It is clear that coordination between government ministries or departments can also be problematic, particularly where there is a mix of deconcentrated and decentralised entities (e.g. **Ethiopia**, **Zimbabwe** and **Ghana**). In **India** while there is a degree of influence from national (federal) level in terms of capacity building and guidelines coming from the DDWS, the main coordination issue is to do with inter-sectoral and inter-ministerial relations. In many states there are still different players in the broader water development and financing arena, with the resultant duplication and lack of coherence between investment schemes, as the example from Tamil Nadu shows in Box 11. In Andhra Pradesh there is a move to introduce an internal or intra-government 'SWAp' to tackle precisely these types of coordination issues.

Discussion: supporting the Service Delivery Approach

Clearly defined policy, institutional, legislative and financial frameworks at national level are a critical foundation for building sustainable services on the ground. Service providers cannot work effectively at the water system level unless everyone knows the 'rules of the game', unless different service levels are agreed upon, and unless it is clear who is responsible to pay for what over the life-cycle of the system. Equally, it is very difficult to support and sustain water supply systems on the ground if there is a free-for-all in terms of different and competing approaches to planning, design and financing, including subsidies.

As with the status of the service delivery authorities, the case studies show a variety of progress towards achieving this enabling environment at national level; this is not surprising given the range of countries included in the study and their relative sector development. Nonetheless, a number of important trends can be identified from the case study contexts.

Financing

Firstly, in almost all cases, it is clear that capital investment costs and minor operation costs are relatively well defined within national policy and macro-level financial planning, although in practice the latter is not always covered by user tariffs, especially in small-scale systems. Very few countries specify the financing requirements for two critical components; namely large-scale CapManEx and replacement expenditure and ExpDS and ExpIDS, including the vital function of post-construction support and monitoring. In cases where this is done in a more systematic way such as **South Africa** and **Uganda**, the resources made available are often insufficient to deal with the maintenance backlog. Asset management planning, which is a relatively common tool for urban utilities, is practically unknown in the rural water sector.

At present many figures for sector investment requirements are estimates, which may not take into account all unit costs, or all possible sources of financing. A key challenge ahead for the adoption of an SDA is for the definition of clearer financial frameworks at sector level that will allow for a much more precise understanding of the expenditure necessary to deliver a service at scale. In absence of these, it is likely that the crucial CapManEx and CapDS will either not be met, or will only be met in an *ad hoc* way, which will simply perpetuate the cycle of failing infrastructure, which is a situation that even wealthier developed countries such as the **USA** can hardly afford.

Capacity support

A critical part of ensuring service provision lies with the back-stopping and support to local government, which under decentralised contexts is often the main guarantor for service provision in its administrative area of jurisdiction. Critical functions of planning, letting of service contracts and monitoring of local operators (whether community-management entities or private sector providers) are increasingly falling on the shoulders of local government. Problems arise where these functions and responsibilities are not always clear to all actors and the reality is that local government capacity is frequently limited. This capacity problem is a critical one, and goes beyond the confines of the water or sanitation sector, being part of a much broader set of issues around public administration delivery. Again, the case studies show a spectrum of experience with support to decentralised government, with more advanced cases of decentralisation showing more systematic support to local government in the area of water (such as **South Africa**, **Uganda** and **India**) than countries with more mixed progress (such as **Mozambique** and **Burkina Faso**).

Aid effectiveness

Increasingly, bi-lateral donors and multi-lateral financing institutions are working within such national sector frameworks, and are moving towards alignment of investment support. But the picture is still mixed, especially for the rural sector; there are cases where donors, INGOs and charities continue to work outside of national frameworks. The negative impact of such continued fragmentation depends on the level of aid dependency, the strength of government vision and leadership, and the relative mix of external donor and NGO activity. What is clear from the studies is that greater levels of harmonisation and alignment – often involving SWAp or sector basket funding mechanisms – support the ability of governments to make at scale, or systemic, investments in strengthening critical systems and structures.

One common concern with the increase in the trend towards harmonisation and SWAps is the trade-off between harmonised approaches and the continued ability for innovation in new ways of working (e.g. testing new handpump designs at scale that may not be sector standard in a particular country, or innovative micro-financing approaches, as was the case with CDF in **Ethiopia**). Put simply, there is an inescapable tension between improved national coordination and alignment by development partners to allow for economies of scale, and the desire to promote local autonomy, experimentation and innovative learning (particularly the role for NGOs). Creating space within SWAps, or whatever sector coordination mechanism exists, to allow and even encourage innovation is clearly important, and may be addressed through funding of research and piloting.

Endnotes

1. Even though the JMP uses one definition for expressing its statistics for all countries, most countries also have their own indicators and statistics for measuring access to rural water supply. Because of differences in definition, often the values are different from the JMP values.
2. The VLOM consists of technologies and systems that are purposefully designed to require minimal external inputs, but primarily refers to handpumps.
3. For further detail on WASMO see: http://www.wasmo.org/default.aspx.
4. For further information on the Uganda Joint Sector Review process and outputs see:
http://www.mwe.go.ug/MoWE/85/Sector_Reviews/Joint_Sector_Review_2010.
5. For more information see: www.ddws.nic.in
6. Mutualisation is a French term referring to the pooling of resources by members of a collective, for the use of members of the collective when they need it. It is used also in banking and insurance. In this case it refers to the pooling of resources for asset management of different water systems in different stages of their life-cycle.
7. Curiously, in some of the countries where there is less or no aid, there is no single sector budget, and investments in the sector tend to come from a myriad of government sources, such as in Colombia and the USA.
8. A number of well-established mechanisms exist at global level such as the International Health Partnership, the Education for All Fast-Track Initiative and the Catalytic Fund.

CHAPTER 5
Conclusions and recommendations

This study seeks to deepen our collective understanding of the status of rural water supply in a range of different country contexts. More specifically, it attempts to identify the factors that contribute to, or constrain, the delivery of sustainable rural water services at scale. Each case country context is distinct, and represents a unique combination of cultural and political history, economic development and aid dependency, water resources, topography and demographic aspects, all of which will have determined the way the rural water sector has evolved. However, there are a number of broad conclusions that can be drawn, and trends that can be highlighted, which appear to cut across these different country experiences. This section looks firstly at an emerging classification of sector development within the context of the SDA. It then examines the factors which appear to have contributed to these scenarios. Finally, it closes with a set of recommendations for moving from a short-term, infrastructure focus to a long-term, service delivery orientation for rural water.

A spectrum of approaches to rural water supply

The study has revealed a spectrum of approaches to rural water supply. We can classify these on a continuum from what can be identified as a largely infrastructure or 'implementation' focus at one extreme, to more SDAs at the other. In simple terms, the former typifies sectors in which structures, systems and efforts are mainly geared towards the capital-intensive phase of rural water (i.e. constructing facilities and initial implementation and training), whereas the latter is geared towards a more balanced attention to the full life-cycle of a service, including aspects such as post-construction support, investment planning for longer-term capital maintenance and asset renewal, and a learning and adaptive sector.

Another way to classify these approaches is according to their degree of scalability. This implies that policy, planning, investment decisions and capacity building is done at a sector (or systemic) level, with the goal of defining clear frameworks and mechanisms for all actors to function within. This type of scalability is associated with those countries that have a relatively higher degree of harmonisation and alignment, and strong decentralised systems. The converse is a situation where sectors are fragmented, investments are made in a 'projectised' way, and structural support for improved capacity is either weak or not available. This can result in poorly defined policy, gaps in legislation, an absence of structured support to decentralised level actors, and confusion about who is responsible for what, particularly regarding financing of different types of expenditure. Not surprisingly, in the latter case, the resultant vacuum tends to be filled by a plethora of 'implementers', many of whom rush to follow their own approaches and policies, simply reinforcing the confusion and lack of commonly agreed 'rules of the game'.

This continuum is summarised in Figure 12 (p141), which indicates some of the more typical characteristics of each stage. Of course, this type of classification is generic in nature and represents extremes; whereas in reality individual country sectors often present a mixed picture, with different elements of these scenarios. For example, in countries such as the **USA** which is not aid dependent and has systems and structures that can work at scale, there are still pressing issues around the need to address the replacement of assets of aging rural water systems. Likewise, **India** relies largely on sovereign financing for the rural sector, and has a relatively advanced level of decentralisation in place, but still faces challenges in many states where the entire system is geared towards construction, and not much is (yet) being done to address the growing problem of slippage in a systematic way.

When viewing the case study countries in this type of classification a number of trends emerge. Firstly, it is apparent that the least developed countries, where current coverage levels are lowest, still mainly adopt what can be termed as an implementation approach; for example, **Ethiopia**, **Mozambique** and **Burkina Faso** all seem to fit in within this pattern. This is logical from the perspective of the political and development imperative facing such countries, being in what Wester (2008) calls the 'hydraulic mission' phase of water development. The focus is first on massive infrastructure development, and ensuring that a sizeable percentage of the population has some form of access to basic services.

Secondly, it seems that as countries achieve higher levels of coverage, a shift is taking place towards adoption of more service-oriented approaches; by strengthening and professionalising rural service providers, putting in place some form of regulation and adopting sustainability and performance indicators. Curiously, this is not always done in a very harmonised way. For example, **Colombia** and the **USA** show a myriad of initiatives for post-construction support, capacity support, and regulation, and a large number of different government agencies and financing streams are involved which are not necessarily harmonised. One possible explanation for this is that these countries are not aid dependent, so efforts to move towards more sector-wide approaches might not have been promoted in the absence of donors concerned with 'aid effectiveness'. On the other hand, this also leads to a certain degree of redundancy in the sense that if one channel for support does not work, others may.

The main difference within this group of countries lies in the extent to which the implementation efforts allow for reaching scale. **Thailand**, **South Africa** and **Uganda** stand out for their scalable approaches. In the case of **Uganda**, for example, there is a strong national policy framework supported by a SWAp, an advanced form of decentralisation, and a certain level of capacity support to local government who fulfil key service authority functions. This situation is reflected in the relatively high rates for functionality seen in the sector, which stood at 81% as reported by the DWD in the last Joint Sector Review (MWE/DWD, 2010). Even accounting for doubts about the accuracy of these figures, the fact that in **Uganda** there are nationally agreed targets and relatively robust structures and systems in place for the rural sector, means that working at scale, and improving the systemic carrying capacity of the sector, is possible in the first place. This compares favourably with other countries where such frameworks do not yet exist (e.g. **Mozambique**) or are still highly fragmented (e.g. **Honduras**), which makes working to improve the sector at scale in the same way much more challenging.

A further remark is that none of the study countries can be said to really have adopted a fully scaled-up SDA. Probably **South Africa** comes closest to this in that it has set up many of the structures and systems which would allow for this approach. However, so far it has largely focused on rapidly increasing coverage through implementation and rehabilitation, with a resultant capital maintenance backlog, despite the fact that it potentially has the systems to start addressing the full life-cycle of services in a more structural way. This in part also reflects the *political nature* of decision-making in the sector, where there is always (more) pressure to allocate resources to new systems, which is the politically more expedient decision,

FROM IMPLEMENTATION OF PROJECTS TO DELIVERY OF SUSTAINABLE SERVICES AT SCALE

Implementation Approach: Limited ability to scale up; time and spatial dimensions are limited	• Focus on interventions through projects at community level • Different management models supported without common agreement • Implementation of parallel and largely uncoordinated programmes with little involvement of decentralised government • Efforts nearly exclusively go into the implementation of new systems or rehabilitation • Planning focuses on implementation of new systems • Monitoring focuses on outputs (systems built and beneficiaries) • Sector targets are defined largely in terms of coverage • Financing mechanisms focus on construction and initial implementation • No systematic support to professionalising service providers
Scaled-up Implementation Approach: May be taken to scale, but does not address long-term systemic change or sustainability of services	• Interventions planned and implemented at scale through coordinated programmes • Involvement of decentralised authorities in implementation • Efforts nearly exclusively go into the implementation of new systems or rehabilitation • Support to skills and capacity building, but limited to implementation only and not to full life-cycle • Monitoring focuses on outputs (systems built and beneficiaries) • Sector targets are defined largely in terms of coverage • Financing mechanisms limited to construction and initial implementation • No systematic support to professionalising service providers
Service Delivery Approach with limited ability to scale up: Supports indefinite services through improving sector systems, but done in a piecemeal way	• Interventions carried out on a project or piecemeal basis with significant gaps • Efforts go into both implementation of new systems and rehabilitation, and to address full life-cycle of a service delivery • Involvement of decentralised authorities in implementation and post-construction • Support to skills and capacity in functions such as planning and regulation; post-construction support starts to address full life-cycle requirements • Monitoring addresses not only outputs, but includes service provided and performance of service providers • Sector targets explicitly include sustainability • Financing mechanisms in place to support capital maintenance and asset management and replacement • Systematic support provided to professionalising service providers • Enabling environment functions of policy and legislation are strengthened • Includes space for technological innovation and testing

Full Service Delivery Approach: Addresses sustainable services at scale through support to entire sector 'system' in a coordinated and comprehensive way	• Interventions planned and implemented at scale through provision of commonly agreed service levels and models for delivery (public, private, etc.) • Efforts address full life-cycle of a service delivery from construction to post-construction, asset management and replacement • Involvement of decentralised authorities in planning, implementation, post-construction and oversight • Support to skills and capacity in functions such as planning and regulation; post-construction support starts to address full life-cycle requirements • Monitoring addresses not only outputs, but includes service provided and performance of service providers • Sector targets explicitly include sustainability • Financing mechanisms in place to support capital maintenance and asset management and replacement • Systematic support provided to professionalising service providers • Enabling environment supports common definitions and frameworks for WASH services; sector learning, policy development, and innovation in technology, financing, etc. is recognised and promoted

■ ■ ■ **Figure 12:** The continuum from implementation towards Service Delivery Approaches

than to spend public or donor funding on the rather invisible tasks of clarifying policy and legislation, or on post-construction support. A number of states in **India** have managed to scale up approaches to rural water provision, most notably the case of WASMO in Gujarat, which has achieved a true level of scale both in terms of physical coverage (upwards of 26 million people) and in creating self-sustaining systems to support CBM at state and district levels.

In spite of this mixed picture in terms of progress, what can be concluded is that there is a general shift, or attempts to shift, towards the SDA. There are many elements of this shift being manifested in the various case studies (see Recommendations section, p147), but most significantly this trend marks the establishment of *sector wide* frameworks that transcend any one (donor) programme, and establish clear roles and legal frameworks. In this way, a multiplicity of actors can be enabled to work within known 'rules of the game' in such a way as to support sustainable provision of a service of a given type and level. This is the logical conclusion of working at scale. This does not necessarily imply a one-size fits all 'national programme'; rather, by providing clear policies, guidelines and norms, a different and appropriate mix of management options can be adopted by different agencies to meet the demands of a range of populations and service levels.

Factors in moving towards the Service Delivery Approach

Recognising the wide range of experiences thrown up by the different case studies, and the fact that in many cases elements of the existing implementation paradigm remain in place, it can be concluded that the rural water supply sector in some countries is now 'maturing'. As part of these processes of reform there are a number of factors that appear to be common in the evolution of sector towards a more service-oriented approach (see Figure 13). Some are more incipient than others; some are still only changes in discourse at this stage, and not yet changes in practice, and there are obvious differences between countries. The most salient factors appear to include the following points made in Figure 13.

1. **Professionalisation of community management**

   Over the past five to ten years there has been a change in the discourse on rural water supply, with an increasing recognition of the limitations of the previous approaches which were largely centred on delivering new infrastructure through projects, with nominal reliance on CBM. As part of this change in discourse there is a growing call for improving CBM, for professionalising aspects that have been traditionally based on voluntary approaches, and for adopting a wider range of service provider options, including private operators.

   A first response to this change in the discourse has been a greater attention to the formalisation of CBM. In many countries the sector has moved to formally recognise CBM in policies and legal frameworks. This has gone

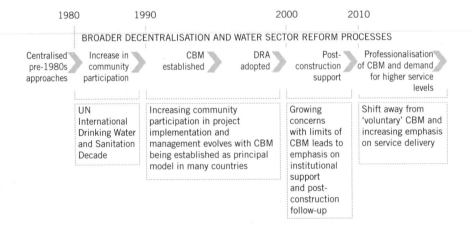

■ ■ ■ **Figure 13:** The evolution of the rural water sector

hand in hand with the *professionalisation* of CBM. This process does not automatically imply turning CBOs into small utilities and privatising them, but rather making them more viable and legally recognised, improving oversight, and moving from voluntary to technically more competent ways of working, including the adoption of more formal contracting. The need for more professional management capacity is also directly correlated to technology options and selection. This has been referred to by some in recent sector dialogue as 'community-management plus' (Moriarty and Verdemato, 2010). However, this is not yet widespread thinking, and traditional, volunteer-based water committees will continue to be the norm particularly around small, point source systems and in more isolated communities. As with previous calls to improve CBM by moving from 'sense of ownership' to 'full legal ownership of assets' (Moriarty and Schouten , 2003), it may now actually be more effective for CBM to make the move from a 'sense of being a service provider' to 'legally being a service provider'.

2. **Increased recognition and promotion of alternative service provider options**

   There is also evidence of a significant and growing role for small-scale *local private operators* in the management of service delivery for some segments of the rural population. Population growth and higher density rural growth centres are making those of the distinction between the demands and solutions of these populations and the truly low-density rural villages and hamlets, increasingly clear. As technological sophistication increases, and there is a move up the ladder from handpumps to reticulated supplies, there will be a need for more specialist operators. Higher levels of service, better and more competent management, and increased opportunities for revenue collection all point towards a more professionalised service. These approaches are more applicable for what can be considered the boundaries between CBM approaches (for smaller, less complex point sources) and delegated models in small towns, rural growth centres, or even for area-based contracts including a larger number of settlements. A second and increasingly formalised option is for *self-supply* in highly dispersed rural areas. The formal recognition of these models in sector policy allows for a more informed selection of different options for different service levels. As the sector moves towards a more service-oriented perspective, the necessary frameworks and mechanisms for enabling private operators to function are being established; these include the legislation, monitoring (and, in certain cases, regulatory frameworks), capacity and business support programmes.

3. **Sustainability indicators and targets**

   The studies show that more and more countries have started collecting information on sustainability and performance indicators of rural water supply services, moving beyond the traditional coverage indicator. This is a sign of growing preoccupation with the actual service users receive, and not only the outputs of implementation projects (e.g. numbers of systems built). Although only a small minority of the study countries have adopted explicit targets for sustainability at sector level, it can be expected that this trend will increase in the coming years. Recognising and formally embedding indicators and targets that seek to measure the level and quality of service provided is a major milestone in the shift towards a more service-oriented way of working.

4. **Standardisation of implementation approaches**

   Many years of practice and research, particularly on CBM approaches, have yielded in-depth insights into key factors during the capital-intensive projects in the life-cycle. The standardisation of implementation allows harmonising and codifying this knowledge in the form of manuals, guidelines and training courses, which in turn allow for obtaining a certain level of quality in these processes as the basis – although not a guarantee – for more sustainable interventions. Alongside the harmonisation of approaches, it is important that opportunities are provided for innovation and learning to take place.

5. **Post-construction support to service providers**

   There is now widespread recognition of the importance of post-construction support and significant experience with the development of models and mechanisms for providing such support. This function is, in fact, increasingly seen as part of professionalising community-based management; such support is also often required to build the capacity of small local private operators. Yet, in spite of this recognition, relatively few countries have moved forward in implementing (and financing) large-scale systems for post-construction support in a structured and sustainable manner.

6. **Capacity support to decentralised government (service authorities)**

   The countries illustrate a wide spread in the depth and speed of decentralisation processes and in sector reform. But, unsurprisingly, the key factor affecting the extent to which local authorities can take up their service authority functions is the capacity of staff and associated systems for procurement, management and monitoring. Many countries have recognised this need, and a number of mechanisms are in place to professionalise local government capacity through support, including through technical assistance from higher-level authorities and association of local governments. Much more can and should be done to provide support to local government, and this support needs to be adequately financed and part of a structured package, rather than relying on project-based support from individual donors or programmes.

7. **Learning and sharing of experience**

   One of the key lessons from the case studies is that a sector which can provide the space and mechanisms to share experiences – both positive and negative – and to learn about what works and what does not work, is generally a more robust and healthy sector. Learning platforms, whether formally endorsed by government or driven by civil society (or ideally including a combination of actors), can provide a critical platform to examine topical issues, to inform policy development, and to undertake trials and innovations. Of course, such mechanisms come with a cost, and these should all be viewed as an integral part of sector expenditure on indirect support.

8. **Political engagement**

    One factor, or driver, that we see consistently across almost all of the country case studies is what can be termed the political economy of rural water. By this we refer to the politics of donor power, of government decision-making and political interference – both from the centre and in the local politics of decentralised government – and its impact on the relative attention to public, private and community-based services providers; choices on resource allocation; investment priorities; as well as issues such as corruption and nepotism. Across the many different contexts the formal development of the rural water sector takes place against a complex backdrop of powerful interests, competing agendas and dynamics, many of which are never formally captured in sector documentation or evaluations.

    In many cases political pressure is manifested in negative terms – the local politician who uses the promise of new water infrastructure to gain votes, or the contractor who pays for the capital cost contribution of communities in order to win the tender. But it is also the donor or NGO that wants to be 'visible' and so ignores, or works outside of, government-led processes. Taken on aggregate, these forces can often reinforce an emphasis on capital investment (in water systems) for financial or political gain, and conversely undermine an emphasis on SDAs which are less expedient.

    But political engagement can also be used for good; this is the identification and nurturing of champions, both politicians and senior and influential civil servants, who can help drive through complex reforms; the establishment of open dialogues between governments and donor agencies to move towards harmonisation; or the possibility to discuss the merits and demerits of different SDMs in line with the predominant political direction of the country.

In addition to these more positive findings, the case studies also highlight a number of gaps or areas of weakness which, if not adequately addressed, will continue to constrain the ability of a sector to move towards the adoption of a more service-oriented approach. These limitations include:

1. **Planning for asset management**

   Investment planning remains one of the critical weaknesses of the rural water supply sector. Currently it is by and large limited to planning for capital investments and rehabilitation only. Structured planning for asset management and renewal is largely absent; the rural equivalent of what utilities would call an Asset Management Plan is largely unknown. As a result, such asset renewal, through rehabilitation, major repairs and replacements, mostly happens in an *ad hoc* manner, often by making financial provision available at the last moment when the need arises; or, more often than not, it does not happen at all, and facilities become non-functional. One of the root causes for this is the poor articulation of CapManEx in financial frameworks (see point 2 below). In addition, there is often a lack of clarity about the level at which such planning could or should take place (e.g. at the level of each individual service provider, the service authority, or at national levels). A possible complicating factor lies with the legal ownership of assets. This tends to lie with either national or local government, but hardly ever do local government agencies feel a sense of ownership of assets that are managed by CBOs; and the latter may not be willing to invest in assets that are not theirs.

2. **Adequate frameworks for financial planning to cover all life-cycle costs**

   This gap is closely related to the previous point, but goes further as it covers a number of specific expenditures that are currently either ignored or not properly accounted for. Most existing financial frameworks define clear responsibilities for investments in CapEx and OpEx, with the former mainly coming from taxes and transfers, and the latter nearly exclusively from tariffs paid by users. However, in most cases, the difference between OpEx (for more minor repairs and operating costs) and more major CapManEx is often ill defined, and there is lack of clarity about who should meet such costs. As a result, the bill for capital maintenance often ends up with (local) authorities, or it is covered in an *ad hoc* manner through other sources of funds. But, as mentioned above, it is hardly ever planned for, which in fact may increase the size of the eventual bill.

   Secondly, financing for the provision of *direct* and *indirect support* functions, which are equally at the heart of sustainable service delivery, are often ill defined. ExpDS is nominally funded through taxes, even though there is scope for user contributions. But in many cases, it is a highly under-funded expenditure category, or it is not clearly defined at all. Post-construction support, backstopping to communities, similar support to local government, and learning platforms are all key to making water flow reliably and continuously from physical infrastructure in communities. It has taken the sector a decade or more to come to this realisation, but the definition of what these functions cost is still unclear. Such lack of clarity

of all cost categories in the life-cycle of services makes it difficult to give a realistic assessment of the adequacy of funding levels to meet sector targets, as well as to be able to direct investments to areas of priority need. Without such comprehensive financial frameworks it will remain very difficult to really adopt an SDA which covers the entire life-cycle of a service.
3. **Regulation of rural services and service providers**
Recognising that rural operators are, or should be, professional service providers implies the need to regulate them; both to protect consumers and to ensure a good quality of service. Present experience with regulation and the associated regulatory functions and actors is very limited. But learning from some early failures, the main lesson appears to be not to adopt a purely punitive perspective of fining rural operators in cases of non-compliance. This has been the case when regulatory frameworks have been translated from urban sectors without a more fundamental re-assessment of the rural context. Rather, it is more useful and appropriate as a mechanism to strengthen the long arm of accountability: holding them accountable for the service they provide. In addition, it is a way to strengthen the performance of service providers; by setting clear rules and standards for the *service provider* and the *service provided* it is possible to establish oversight, track performance against the standards, and direct efforts to improve performance. In such a way regulation can, in fact, contribute to or encourage improved performance rather than taking a punitive approach.

Recommendations for the sustainable provision of rural water services at scale

We all share a vision to see better and more reliable water services supplied to rural populations, and to see the benefits of investments sustained over as long a time period as possible. But what recommendations can we give that can promote the adoption of more service-orientated approaches that can work at scale?

The first remark is that these recommendations should be country-specific. As seen throughout this book, different countries currently follow different approaches, but in almost all cases there are signs of progress – some more advanced than others – in the adoption of elements of a more service-orientated approach. We therefore have to differentiate between countries which are struggling to meet the basic challenge of increasing coverage, and those that are moving towards so-called 'second generation' problems once a critical mass of coverage has been achieved. In considering the policy implications for how to tackle this problem, we can think of three broad sets of countries:
- Firstly, those in which coverage is really still very low (for example, **Ethiopia** and **Mozambique**). It is an understandable strategy to focus largely on increasing coverage, but this should be, as far as possible, in

a scaled-up manner. While much energy and resources go into building new systems, there may have to be an acceptance of a case of 'two steps forward, one step back', where levels of functionality will remain problematic. Where CBM is the main SDM, it should be strengthened, and incorporate changes flagged by this book, particularly in the legal recognition of committees and formalising their relationship with local government. Post-construction support must be adequately addressed, particularly for this group this is one element that has been consistently under-funded in the past. Development partner assistance to countries in this group should focus on improving alignment of programmatic support, particularly around implementation approaches, to avoid fragmentation and conflicting policies for communities.

- Secondly, we have a group of countries where coverage is already relatively high, reaching levels of 80% or more (for example, **India**, **Thailand**, the **USA** and **Sri Lanka**) which should focus very strongly on investing in systems and capacities that underpin a true SDA. Such steps would include developing asset management planning, providing structured capacity support to local government, post-construction support and financial mechanisms such as rotating funds to meet CapManEx, improving life-cycle cost analysis and regulation. Another important step would be to develop specific strategies to reach the last 10-15% of the unserved populations, for example, by formally recognising self-supply, and introducing measures to support this approach in a systematic way. Developing asset management strategies and tools for rural water services would mark another major step-change in these countries.

- Lastly, we have the middle band of countries where coverage is somewhere between 50% and 70% and expanding, but where there is also the very strong risk of slippage of functionality rates (for example, **Honduras**, **Colombia**, **Ghana**, **Uganda** and **Burkina Faso**). These countries face an in-built tension between pursuing increased coverage (with inadequate budgets and growing populations), while at the same time addressing sustainability in a more structured way. More capital investment is needed in new systems, extensions are needed to existing systems, but, equally, increasing attention needs to be paid to asset management, improving management options, and monitoring and oversight of services delivered. So how should this group of countries juggle all of these balls at once?

This is indeed a critical question and the simple answer is that these middle groups must juggle competing priorities as part of this phase of sector development. We cannot give an estimate of how much should go into new investments and how much into asset management or, in other words, what is required to build (new) systems and what is required to build and maintain the sector capacity to support the delivery of services, but what is clear is that for many countries the balance has to shift in favour of the latter. This should also be a strong message for development partners when considering funding strategies. Having a robust scaled-up implementation approach (such as the sectors adopted in **South Africa**, **Uganda** and **India** did when they were themselves only climbing the lower levels of coverage) has helped in making the transition to an SDA in subsequent phases.

Both national governments and development partners should invest more in building the systems of the sector to cope with the transition to service delivery, including support to professionalising CBM, capacity support to decentralised government sector staff, and clarifying the legal and institutional frameworks for asset management and delegated contracting. Setting up and streamlining financial mechanisms and the introduction of pooled funding would allow for support to these type of efforts to improve the carrying capacity of the sector. Overall, development partners should take the long-view with this group of countries and move from the two to three year support horizons to much longer-term, more stable funding support which will allow for sector development in a more predictable way. Conversely, governments must commit to following through with and supporting sector reforms that often exist on paper only.

The recommendations to sector policy makers and practitioners are to focus on the range of factors that are already emerging as important in the shift towards an SDA as outlined in the findings from this study, as well as starting to address the three areas of weakness which so far appear not to have been addressed in a substantive way in the majority of cases. Table 14 (p150) summarises these recommendations in a shift from the 'implementation end' of the spectrum towards a more service-oriented perspective that should be adopted by this key group of middle countries.

Implications for sector change

Finally, the country studies highlight three very important underlying lessons for promoting more service-oriented outcomes; these can be considered as broad filters or lenses through which to assess efforts in improving sector performance.

The first is the *tension between broad-based systemic change and gradual improvements* in specific areas. Sector policy, institutions, legislation and structures must be clarified and modified, as necessary, to enable the support of a service rather than the delivery of infrastructure. The result is a clear allocation of roles and functions to different actors, supported by clear legal frameworks in such a way as to support sustainable provision of a service of a given type and level. The 'rules of the game' must be clear to all players working at all levels, from ministry and local government staff, to donor partners, water committee members, private operators and consumers. To achieve real change the entire system needs to be addressed. Attempting to make changes through

Table 14: Recommendations for shifting from implementation to Service Delivery Approaches

Factors	From: implementation approaches with limited scale	To: more sustainable services delivered at scale
Professionalisation of community management	Community management based on voluntary principles and without legalisation or clear contracting	Community management properly embedded in, and supported by, policy, legal, regulatory frameworks and support services
Increased recognition and promotion of alternative service provider options	Community-based management taken as *de facto* management option	Range of management options set out in clear SDMs that are differentiated by service levels, technology and types of settlements
Sustainability indicators and targets	Coverage as the principal sector monitoring benchmark and target	Monitoring and target indicators that benchmark against services delivered and performance of service providers
Standardisation of implementation approaches	Project-based or organisational implementation manuals which are duplicative or conflicting, and vary according to funder and implementer	Sectoral implementation manuals which set out common frameworks, norms and standards, but with flexibility in implementation and space for innovation
Post-construction support to service providers	Water committees set up during implementation processes left to manage without structured follow-up support	Structured system of support established and funded to back-up and monitor community management entities or small private operators
Capacity support to decentralised government (service authorities)	Local authorities receive *ad hoc* support from projects during implementation, and are left to manage without support after systems are built	Ongoing capacity support programme to local authorities covering key functions in the life-cycle of rural water supply services including management, procurement and contracting
Learning and sharing of experience	Learning and knowledge management happens in an *ad hoc* way or as a by product of implementation, and is not funded as a stand alone investment	Learning and knowledge management are taken as integral parts of sector capacity, and supported at national and decentralised levels
Planning for asset management	Planning focuses on implementation of new systems and/or their rehabilitation	Asset management planning is carried out systematically, with financial forecasting and inventory updates for the entire life-cycle
Financial planning frameworks to cover all life-cycle costs	Financial frameworks specify rules around funding sources for CapEx and OpEx	Sector financial frameworks consider full life-cycle costs, especially CapManEx, ExDS and ExIDS
Regulation of rural services and service providers	Oversight over implementation processes during construction and preparation of communities	Regulation of rural water supply services and service providers through appropriate mechanisms/ regulatory agents at the local level

isolated projects and programmes, by setting up 'stand-alone' solutions, will have limited impact. Yet, achieving systemic change is not straightforward and, in many cases, one has to start pragmatically addressing one or more of the building blocks presented.

In order to help maintain the balance between the overall picture and the specific areas of priority action, it can be useful to use frameworks such as the one we have used and adapted for this study. It is intended to be used as a guide to allow for a gap analysis at different levels at scale, and for planning of interventions or changes in different aspects of improving service delivery (see Annex D). Based on the experiences in this study, we feel it has been useful as a framework to identify specific actions for improvement while keeping the inter-connectivity of different elements, thereby avoiding the trap of reducing sustainability issues to one or two key factors.

Secondly, in order to achieve systemic change there must be a *base level of harmonisation and coordination* between different actors working in the sector; this is particularly the case for the more aid-dependent countries, but is also a factor in intra-government relationships. Common agreement and adherence to sector policy, norms and guidelines is an essential building block for working at scale. The last decade has seen a steady rise of more harmonised and aligned approaches, particularly the adoption of coordinated implementation approaches, SWAps, basket funding, sector coordination platforms, joint performance reports, etc. Working through these more harmonised structures is also one of the best ways to address and finance systemic capacity building.

Finally, there is the issue of *the political economy of rural water*. In order to achieve change one has to recognise that changes in approaches to rural water reflect profound political choices, and that one has to embrace the entire spectrum of politics. We therefore would argue in favour of change processes which are strongly vested in the political agendas, both nationally and locally, of all actors involved. Change processes therefore need to be accompanied by and embedded in political engagement activities.

References

Abrams, L., Palmer, I., and Hart, T. 1998. *Sustainability Management Guidelines*. Pretoria: Department of Water Affairs and Forestry.

Adjinacou, C., 2011. *Benin: L'alimentation en eau en zones rurales; evaluation des progrès vers la prestation de services durables*. Unpublished report for IRC International Water and Sanitation Centre prepared by MGE Conseils.

AGUASAN, 2008. *Promising management models of rural water supply services. Outcomes of 24th Workshop. Gwatt, Switzerland, 13 to 17 October, 2008*. AGUASAN Workshop Series. Berne, Switzerland: SDC, Eawag/Sandec, Helvetas and Skat.

Asian Development Bank, 2010. Sri Lanka Fact Sheet. Manila: ADB. Available at http://www.adb.org/Documents/Fact_Sheets/SRI.pdf

Asthana, A.N., 2003. Decentralisation and Supply Efficiency: The case of Rural Water Supply in Central India. *Journal of Development Studies*, 39 (4), pp 148-159.

Azuba, C., Mugabi, J. and Mumssen, Y., 2010. *Output based aid for water supply in Uganda: increasing access in small towns*, OBA Approaches, Note number 35, July 2010. Washington, D.C.: World Bank/Global Partnership on Output-Based Aid.

Bakalian, A. and Wakeman, W., 2009. *Post-construction support and sustainability in community-managed rural water supply: case studies in Peru, Bolivia and Ghana*, Working Paper Report, Number 48731, 1 (1), Water Sector Board discussion paper series, no. 14 Washington, D.C.: Bank-Netherlands Water Partnership (BNWP) and World Bank.

Barraqué, B., 2009. The Development of Water Services in Europe: from Diversity to Convergence. In: Castro, J.E. and Heller, L., eds., 2009. *Water and sanitation services; public policy and management*. London: Earthscan.

Brussee, P., Marín, X. y Smits, S., 2010. *Memoria del seminario de intercambio de experiencias; gobernanza de servicios de saneamiento sostenibles en Centroamérica*. San Salvador, 1-3 de Febrero 2010. The Hague: IRC International Water and Sanitation Centre [online] Available at: http://www.es.irc.nl/page/52808.

Butterworth, J.B., 2010. *A brief review of service delivery concepts; a literature review*. The Hague: IRC International Water and Sanitation Centre.

Butterworth, J., Warner, J., Moriarty, P., Smits, S. and Batchelor, C., 2010. Finding practical approaches to Integrated Water Resources Management. *Water Alternatives*, 3 (1), pp. 68-81.

Castro, J.E. and Heller, L., eds., 2009. *Water and sanitation services; public policy and management*. London: Earthscan.

Chaka, T. Yirgu, L. Abebe, Z. and Butterworth, J., 2011. *Ethiopia: Lessons for Rural Water Supply: Assessing progress towards sustainable service delivery*. The Hague: IRC International Water and Sanitation Centre.

Comprehensive Assessment of Water Management in Agriculture, 2007. *Water for Food, Water for Life: A Comprehensive Assessment of Water Management in*

Agriculture. London: Earthscan and Colombo: International Water Management Institute.

Danida, 2006. *Harmonisation and Alignment in water sector programmes and initiatives: Good practice paper*. Copenhagen: Ministry of Foreign Affairs of Denmark, Technical Advisory Services.

DAR/DNA, 2010. *Estratégias e Plano de Manutenção das Fontes – Apresentação ao Conselho Coordenador*. Maputo: *Direcção Nacional de Águas*.

Davis J. and Iyer P., 2002. *Taking Sustainable Rural Water Supply Services to Scale. A Discussion Paper*.Washington, D.C.: Bank Netherlands Water Partnership – Water and Sanitation Program.

DDWS web site: www.ddws.nic.in.

de la Harpe, J., 2006. *Municipal Infrastructure, roles and responsibilities of national, provincial and local government*. Pretoria: Department of Provincial and Local Government.

de la Harpe, J., 2011a. *Harmonisation and alignment literature review*. The Hague: IRC International Water and Sanitation Centre.

de la Harpe, J., 2011b. *South Africa: Lessons for Rural Water Supply; Assessing progress towards sustainable service delivery*. The Hague: IRC International Water and Sanitation Centre.

Evans, P., 1992. *Paying the piper: an overview of community financing of water and sanitation*. Delft: IRC International Water and Sanitation Centre.

FAO, 2007. *Modernizing Irrigation Management – the MASSCOTE approach*. FAO Irrigation and Drainage Paper series 63. Available at: http://www.fao.org/docrep/010/a1114e/a1114e00.htm.

FAO, 2010. *Aquastat database*. [online] Available at: http://www.fao.org/nr/water/aquastat/main/index.stm.

Fonseca, C., 2011a forthcoming. *The challenge of capital maintenance in rural water supply in developing countries: learning from historical (accounting) developments*. The Hague: IRC International Water and Sanitation Centre.

Fonseca, C., 2011b forthcoming. *A review of the literature on unit costs for rural and peri-urban water supply and sanitation in developing countries (2000-2010)*. The Hague: IRC International Water and Sanitation Centre.

Fonseca, C and Diaz, C., 2008. *Working Together to Improve Aid Effectiveness in the Water Sector*. The Hague: European Union Water Initiative, Africa Working Group.

Fonseca, C., Franceys, R., Batchelor, C., McIntyre, P., Klutse, A., Komives, K., Moriarty, P., Naafs, A., Nyarko, K., Pezon, C., Potter, A., Reddy, R. and Snehalatha, M., 2010. *Life-Cycle Costs Approach; Glossary and cost components*. The Hague: IRC International Water and Sanitation Centre.

Gasteyer, S., 2011. *United States of America: Lessons for Rural Water Supply; Assessing progress towards sustainable service delivery*. The Hague: IRC International Water and Sanitation Centre and Michigan, USA: Michigan State University.

Gibson, J., 2010. *Operation and maintenance costs of rural water supply schemes in South Africa*. Paper presented at: *Pumps, Pipes and Promises: Costs, Finances and Accountability for Sustainable WASH Services*, The Hague, 16 - 18 November 2010. The Hague: IRC International Water and Sanitation Centre. [online] Available at: http://www.irc.nl/page/55569.

Godfrey, S., Freitas, M., Muianga, A., Amaro, M., Fernandez, P. and Sousa Mosies, L., 2009. *Sustainability check: A monitoring tool for the sustainability of rural water* supplies. Paper presented at the *34th WEDC International Conference: Water: Sanitation and Hygiene: sustainable development and multisectoral approaches, Addis Ababa*. Loughborough, WEDC.

GoI, 2008. *Movement Towards Ensuring People's Drinking Water Security In Rural India*. Delhi: Government of India.

GSS, GHS and ICF Macro, 2009. *Demographic and Health Survey 2008*. Accra: Ghana Statistical Service, Ghana Health Service and ICF Macro.

Gundel, S.; Hancock, J. and Anderson, S., 2001. *Scaling up strategies for research in natural resources management: a comparative review*. Chatham: Natural Resources Institute.

Hall, D., and Lobina, E., 2010. *The past, present and future of finance for investment in water systems*. Paper presented at: *Pumps, Pipes and Promises: Costs, Finances and Accountability for Sustainable WASH Services*, The Hague, 16-18 November 2010. The Hague: IRC International Water and Sanitation Centre. [online] Available at: http://www.irc.nl/page/55569.

Harrington, L., White, J., Grace, P., Hodson, D., Dewi Hartkamp, A., Vaughan, C. and Meisner, C., 2001. Delivering the goods: scaling out results of natural resource management research. *Conservation ecology*, 5 (2), Art. 19. [online] Available at: http://www.consecol.org/vol5/iss2/art19/.

Harvey, P. and Reed, B., 2004. *Rural Water Supply in Africa; Building Blocks for Handpump Sustainability*. Loughborough: Water, Engineering and Development Centre.

Harvey, P.A. and Reed R.A., 2006. Community-managed water supplies in Africa: sustainable or dispensable? *Community Development Journal*, 42 (3), pp. 365-378.

Hutton, G. and Bartram, J., 2008. Global costs of attaining the Millennium Development Goal for water supply and sanitation. *Bulletin of the World Health Organization*, 86, pp.13–19.

IMF, 2010. *World Economic Outlook Database*. [online] Available at: http://www.imf.org/external/pubs/ft/weo/2010/01/weodata/index.aspx.

IRC/Aguaconsult, 2008. Sustainable Services at Scale project proposal, unpublished.

IRC web site: www.irc.nl.

IRC, 2009. *Sustainable Services at Scale (Triple-S), Briefing Note*, November 2009. The Hague: IRC International Water and Sanitation Centre. [online] Available at: http://www.irc.nl/page/51032.

IRC/Aguaconsult, 2011. *Ghana: Lessons for Rural Water Supply; Assessing progress towards sustainable service delivery*. Accra, Ghana: IRC International Water and Sanitation Centre.

Iyer, P., Davis, J. and Yavuz, E., 2006. *Rural Water Supply, Sanitation, and Hygiene: A Review of 25 Years of World Bank Lending (1978-2003) – Summary Report*. Water Supply and Sanitation Working Notes, Note No. 10, July 2006. Washington, D.C.: The World Bank.

James, A. J., 2004. *India's Sector Reform Projects and Swajaldhara Programme: A Case of Scaling up Community Managed Water Supply*. [online] Available at: http://www.irc.nl/docsearch/title/126451.

James, A.J., 2011a. *India: Lessons for Rural Water Supply; Assessing progress towards sustainable service delivery*. The Hague: IRC International Water and Sanitation Centre and Delhi: iMaCS.

James, A.J., 2011b. *Sri Lanka: Lessons for Rural Water Supply; Assessing progress towards sustainable service delivery*. The Hague: IRC International Water and Sanitation Centre and Delhi: iMaCS.

Juntopas, M. and Naruchaikusol, S., 2011. *Thailand: Lessons for Rural Water Supply; Assessing progress towards sustainable service delivery*. The Hague: IRC International Water and Sanitation Centre and Bangkok: Stockholm Environment Institute, Asia Centre.

KA Associates, 1999. *CWSA Ghana Unit cost report*. Accra: Community Water Supply Agency.

Kayser, G., Griffiths, J., Moomaw, W., Schaffner, J. and Rogers, B., 2010. Assessing the Impact of Post-Construction Support—The Circuit Rider Model–on System Performance and Sustainability in Community Managed Water Supply: Evidence from El Salvador. In: Smits, S., Lockwood, H., Danert, K., Pezon, C., Kabirizi, A., Carter, R. and Rop, R., 2010. *Proceedings of an international symposium*. Kampala, 13-15 April 2010. The Netherlands: Thematic Group on Scaling Up Rural Water Services.

Kleemeier, K., 2008. *Private Provision of Rural Piped Water in Bangladesh* Draft Field Note prepared for Water and Sanitation Program–South Asia, Bangladesh Country Office.

Koestler, L., Koestler, A.G., Koestler, M.A. and Koestler, V.J., 2010. Improving sustainability using incentives for operation and maintenance: The concept of water-person-years. *Waterlines*, 29 (2),pp. 147-162.

Lockwood, H., 2002. *Institutional Support Mechanisms for Community-managed Rural Water Supply and Sanitation Systems in Latin America*, Strategic Report 6, Environmental Health Project (EHP). Washington, DC: USAID. [online] Available at: http://pdf.dec.org/pdf_docs/PNACR786.pdf.

Lockwood, H., 2004. *Scaling Up Community Management of Rural Water Supply*. Thematic Overview Paper. The Hague: IRC International Water and Sanitation Centre.

López, M.A., 2011. *Honduras: Abastecimiento de agua en zonas rurales; Experiencias en la prestación de servicios sostenibles*. The Hague: IRC International Water and Sanitation Centre and Tegucigalpa, Honduras: RASHON.

Makoni, F., Smits, S., Shoshore, I. and M. Jonga 2007. `Institutional learning about multiple use services in Zimbabwe; experiences of the Learning Alliance approach'. MUS Project Working Paper. Harare, Zimbabwe. http://www.musgroup.net/page/1000

Malano, H. and van Hofwegen, P., 1999. *Management of irrigation and drainage systems; a service approach*. IHE Monograph series. Rotterdam: Balkema Publishers.

MAPLE Consult/WSMP, 2010. *Compilation of Information/Data on Water and Sanitation Sector Investments in Ghana*. Report submitted by MAPLE Consult to the WSMP, January 2010. Accra: Water and Sanitation Monitoring Platform.

Maria Solo, T., 2003. *Independent Water Entrepreneurs in Latin America; the other private sector in water services*. Washington, D.C.: The World Bank.

Makoni, F., Smits, S., Shoshore, I. and M. Jonga (2007) *Institutional learning about multiple use services in Zimbabwe; experiences of the Learning Alliance approach*. MUS Project Working Paper. Harare, Zimbabwe. http://www.musgroup.net/page/1000

McGranahan, G. and Mulenga, M., 2009. Community organization and alternative paradigms for improving water and sanitation in deprived settlements. In:

Castro, J.E. and Heller, L., eds., 2009. *Water and sanitation services; public policy and management.* London: Earthscan.
MEE, 2010. *Budget programme 2010-2012 hydraulique en milieu rurale.* Cotonou: Ministere De l'Eau et de l'Energie.
Moriarty, P., Batchelor, C., Fonseca, C., Klutse, A., Naafs, A., Nyarko, A., Pezon, K., Potter, A., Reddy, R. and Snehalata, M., 2010a. *Ladders and levels for assessing and costing water service delivery.* WASHCost working paper No. 2. The Hague: IRC International Water and Sanitation Centre.
Moriarty, P., Butterworth, J. and Batchelor, C., 2004. *Integrated water resources management and the domestic water and sanitation sub-sector.* Thematic overview paper. Delft: IRC International Water and Sanitation Centre. [online] Available at: www.irc.nl/page/10431.
Moriarty, P., Naafs, A., Pezon, C., Fonseca, C., Uandela, A., Potter, A., Batchelor, C., Reddy, R. and Snehalata, M., 2010b. *WASHCost's theory of change; reforms in the water sector and what they mean for the use of unit costs.* WASHCost working paper No. 1. The Hague: IRC International Water and Sanitation Centre.
Moriarty, P. and Schouten, T., 2003. *Community Water, Community Management: from system to service in rural areas.* UK: ITDG Publishing.
Moriarty, P. and Verdemato, T., 2010. *Discussion Report of the International Symposium on Rural Water Services: Providing Sustainable Water Services at Scale,* Kampala, 13-15 April 2010. The Hague: Thematic Group on Scaling Up Rural Water Services [online] Available at: www.scalingup.watsan.net.
Munguambe, C.G.J. and Langa de Jesus, V.A., 2011. *Moçambique: Abastecimento de água nas zonas rurais; Avaliação dos progressos para a prestação de servíços sustentáveis.* The Hague: IRC International Water and Sanitation Centre.
MWE web site: http://www.mwe.go.ug/.
MWE/DWD, 2008a. *Tracking study for the water and sanitation sector (WSS) cost variation.* Kampala: Ministry of Water and Environment, Directorate of Water Development.
MWE/DWD, 2008b. *Water and Sanitation Sector Performance Report.* Kampala: Ministry of Water and Environment, Directorate of Water Development.
MWE/DWD, 2010. *Water and Environment Sector Performance Report 2010.* Kampala: Ministry of Water and Environment, Directorate of Water Development.
Nimanya, C., Nabunnya, H., Kyeyune, S. and Heijnen, H., 2011. *Uganda: Lessons for Rural Water Supply; Assessing progress towards sustainable service delivery.* The Hague: IRC International Water and Sanitation Centre and Kampala: NETWAS.
OECD, 2006. *2005: Development Co-operation Report,* 7 (1). Paris: Organisation for Economic Co-operation and Development.
OECD, 2008a. *Paris Declaration on Aid Effectiveness and Accra Agenda for Action.* Paris: Organisation for Economic Co-operation and Development.
OECD , 2008b. *2008 Survey on monitoring the Paris Declaration: making aid more effective by 2010.* Paris: Organisation for Economic Co-operation and Development. [online] Available at: http://www.oecd.org/dataoecd/58/41/41202121.pdf.
OECD, 2009. *DAC list of ODA recipients.* [online] Available at: http://www.oecd.org/dataoecd/32/40/43540882.pdf.
OECD, 2010. Strategies to improve rural service delivery. *OECD Rural Policy Reviews.* Paris: Organisation for Economic Co-operation and Development.
Office of the Auditor General, 2009. *Value for money audit report on provision of water and maintenance of water facilities in District Local Governments by the Directorate*

of Water Development, Ministry of Water and Environment. Kampala: Government of the Republic of Uganda.

Ofwat, 2005. *Water and sewerage service unit costs and relative efficiency 2003–2004 report*. Birmingham: Water Services Regulation Authority.

Oyo, A., 2006. *Spare Part Supplies for Handpumps in Africa; Success Factors for Sustainability*. Rural Water Supply Series, Field Note. St Gallen: WSP/Rural Water Supply Network.

Pearson, M., 2007. *US Infrastructure Finance Needs for Water and Waste-water*. Washington, D.C.: Rural Community Assistance Partnership. [online] Available at: http://www.map-inc.org/finance.

Pezon, C., 2009. Decentralisation and Delegation of Water and Sanitation Services in France. In: Castro, J.E. and Heller, L., eds., 2009. *Water and sanitation services; public policy and management*. London: Earthscan.

Rivas Hermann, R., Paz Mena, T., Gómez, L.I. and Ravnborg, H., 2010. *Cooperación y Conflicto en torno a la Gestión Local del Agua en el municipio de Condega, Nicaragua*. DIIS Working Paper 2010:13. Copenhagen: Danish Institute for International Studies.

Rivera Garay, C.J. and Godoy Ayestas, J.C., 2004. *Experiencias, Estrategias y Procesos Desarrollados por Honduras en el Sector Agua Potable y Saneamiento en el área Rural*. Foro Centroamericano y Republica Dominicana de Agua Potable y Saneamiento, August 2004.

Rojas, J., Zamora, A., Tamayo, P. and García, M., 2011. *Colombia: Abastecimiento de agua en zonas rurales; Experiencias en la prestación de servicios sostenibles*. The Hague: IRC International Water and Sanitation Centre and Cali: Universidad del Valle/CINARA.

RWSN, 2009. *Myths of the Rural Water Supply Sector*, Perspectives No. 4, RWSN Executive Steering Committee, July 2009. St Gallen: Rural Water Supply Network. [online] Available at: http://www.rwsn.ch.

Ryan, P., 2006. *Citizens' Action for water and sanitation*, Discussion Paper, January 2006. London: Wateraid.

SANAA, 2009. *Sistema de Información de Acueductos Rurales (SIAR) database*. Tegucigalpa: Servicio Autónomo Nacional de Acueductos y Alcantarillados.

Serrano, P., 2007. *Formulación programa de inversiones del sector agua potable y saneamiento*. Tegucigalpa: Consejo Nacional de Agua Potable y Saneamiento/Comisión Presidencial de Modernización del Estado.

SIDA, 2009. *Support to Uganda's Water and Sanitation Sector from the 1980's onwards – reflections and experiences*. Stockholm: Swedish International Development Cooperation Agency.

Skinner, J., 2009. *Where every drops counts: tackling Africa's water crisis*, IIED Briefing paper. [online] Available at: http://www.iied.org/pubs/pdfs/17055IIED.pdf [accessed June 2010].

Smits, S. and Butterworth, J., 2006. *Literature review: Local government and integrated water resources management*, LoGoWater project report. Freiburg: ICLEI. [online] Available at: www.iclei-europe.org/index.php?id=1587.

Smits, S., Lockwood, H., Danert, K., Pezon, C., Kabirizi, A., Carter, R. and Rop R., 2010. *Proceedings of the International Symposium on Rural Water Services: Providing Sustainable Water Services at Scale*, Kampala, 13-15 April 2010. The Netherlands: Thematic Group on Scaling Up Rural Water Services.

Sutton, S., 2007. *An introduction to Self-Supply. Putting the User First: Incremental improvements and private investment in rural water supply*, WSP Field Note, Rural Water Supply Network, Self-Supply Flagship. Nairobi: Water and Sanitation Program.

Tamayo, S.P. y García, M., 2006. Estrategia estatal para el fortalecimiento de entes prestadores de servicios públicos en el pequeño municipio y la zona rural. El programa cultura empresarial adelantado en Colombia. In: Quiroz, F., Faysse, N. y Ampuero, R., 2006. *Apoyo a la gestión de Comités de Agua Potable; experiencias de fortalecimiento a comités de agua potable con gestión comunitaria en Bolivia y Colombia*. Cochabamba: Centro Agua – UMSS.

Taylor, B. (2009). *Addressing the Sustainability Crisis: lessons from research on managing rural water projects*. Dar es Salaam: WaterAid.

Thematic Group for Scaling Up of Community Management for Rural Water Supply, 2005. *Scaling Up Rural Water Supply: A framework for achieving sustainable universal coverage through community management*. The Netherlands: IRC International Water and Sanitation Centre [online] Available at: http://www.scalingup.watsan.net.

Thompson, M., Okuni, P.A. and Sansom K., 2005. *Sector performance reporting in Uganda – from measurement to monitoring and management. Paper presented at 31st WEDC International Conference: Maximising The Benefits From Water And Environmental Sanitation*, Kampala, 2005. Loughborough: WEDC.

UNICEF/NAC, 2004. *WASH Inventory ATLAS Zimbabwe: Inventory of the national rural water supply and sanitation facilities*. Harare: UNICEF/NAC.

van Koppen, B., Moriarty, P. and Boelee, E., 2006. *Multiple-use water services to advance the Millennium Development Goals*, Research report no. 98. Colombo: International Water Management Institute.

van Koppen, B., Smits, S., Moriarty, P., Penning de Vries, F., Mikhail, M. and Boelee, E., 2009. *Climbing the Water Ladder: Multiple-use water services for poverty reduction*, TP series, no. 52. The Hague: IRC International Water and Sanitation Centre and Colombo: International Water Management Institute.

van Wijk-Sijbesma, C.A., 2001. *The best of two worlds?: methodology for participatory assessment of community water services*, Technical paper series/IRC, no. 38. Delft: IRC International Water and Sanitation Centre.

Visscher J.T. and Da Silva Wells, C., 2006. *Landscaping and Review of Approaches to support service provision for Water, Sanitation and Hygiene*. Delft: IRC International Water and Sanitation Centre, London: Aguaconsult and Cranfield: Cranfield University.

WASHCost web site: http://www.washcost.info/.

WASMO web site: http://www.wasmo.org/.

Water Partnership Programme/AfDB, 2009. *Water Sector Governance in Africa, Volume 1: Theory and Practice*. Tunis: African Development Bank.

WaterAid, 2008. *Think local, act local: effective financing of local governments to provide water and sanitation services*. London: WaterAid.

WaterAid, 2009. *Management for Sustainability: Practical lessons from 3 studies on the management of rural water supply schemes*. Dar es Salaam: WaterAid. [online] Available at: http://www.wateraid.org/documents/plugin_documents/management_for_sustainability.pdf.

Welle, K., Nicol, A. and Van Steenbergen, F., 2008. *Why is harmonisation and alignment difficult for donors?: lessons from the water sector*, Project briefings/ODI. London: ODI.

Wester, P., 2008. *Shedding the waters: institutional change and water control in the Lerma-Chapala Basin, Mexico.* Ph. D. Netherlands: Wageningen University.

Whittington, D., Davis, J., Prokopy, L., Komives, K., Thorsten, R., Lukacs, H., Bakalian, A. and Wakeman, W., 2009. How well is the demand-driven, community management model for rural water supply systems doing? Evidence from Bolivia, Peru, and Ghana. In: *Water Policy*, 11, pp. 696-718.

WHO, 2010a *UN-water global annual assessment of sanitation and drinking-water (GLAAS) 2010: targeting resources for better results.* Geneva: World Health Organization.

WHO, 2010b. *Access to improved drinking-water sources and to improved sanitation (percentage).* WHO Statistical Information System website. [online] Available at: http://www.who.int/whosis/indicators/compendium/2008/2wst/en/

WHO/UNICEF, 2010. *Progress on Sanitation and Drinking-water: 2010 Update.* Geneva: WHO/UNICEF Joint Monitoring Programme for Water Supply and Sanitation.

World Bank, 2004a. *Colombia: Desarrollo Económico Reciente en Infraestructura; Balanceando las necesidades sociales y productivas de infraestructura. Informes de Base Sector Agua Potable.* Washington, D.C.: The World Bank.

World Bank, 2004b. *World Development Report 2004: making services work for poor people*, Report No. 26895. Washington, D.C.: The World Bank.

World Bank, 2009. *Implementation Completion and Results Report for the Kerala Rural Water Supply and Environmental Sanitation Project.* Washington, D.C.: The World Bank.

World Bank, 2010. *World Development Indicators.* [online] Available at: http://data.worldbank.org/indicator/DT.ODA.ODAT.GN.ZS.

World Bank/IEG, 2008. *Decentralization in Client Countries. An Evaluation of World Bank Support, 1990–2007.* Washington, D.C.: The World Bank/Independent Evaluation Group.

WSP, 2004. *Decentralization of rural water and sanitation services: New roles for rural water associations and boards in Honduras*, Field Note. Lima: Water and Sanitation Program.

WSP, 2010. *A Practitioners' Workshop Report: Sustainable Management of Small Water Supply Systems in Africa.* Maputo, 6-8 October 2010. Nairobi: Water and Sanitation Program.

Zoungrana, D., 2011. *Burkina Faso: L'alimentation en eau en zones rurales; évaluation des progrès vers la prestation de services durables.* The Hague: IRC International Water and Sanitation Centre, and Ouagadougou: Institut International d'Ingénierie de l'Eau et de l'Environnement.

ANNEX A
List of abbreviations, acronymns and non-english terms

AFD	*Agence Française de Développement* [French Development Agency]
CapEx	Capital Expenditure
CapManEx	Capital Maintenance Expenditure
CBM	Community-Based Management
CBO	Community-Based Organisation
CoC	Cost of Capital
Danida	Danish International Development Assistance
DFID	Department for International Development [United Kingdom]
DRA	Demand-Responsive Approach
EU	European Union
ExpDS	Expenditure on Direct Support
ExpIDS	Expenditure on Indirect Support
GDP	Gross Domestic Product
GIZ	Gesellschaft für Internationale Zusammenarbeit [German international cooperation organisation]
HDI	Human Development Index
INGO	International Non-Governmental Organisation
IRC	International Water and Sanitation Centre [the Netherlands]
IWRM	Integrated Water Resources Management
JMP	Joint Monitoring Program
lpcd	litres per capita per day
MDG	Millennium Development Goal
NGO	Non-Governmental Organisation
O&M	Operations and Maintenance
OECD	Organisation for Economic Co-operation and Development
Ofwat	Water Services Regulation Authority [in England and Wales]
OpEx	Operating and minor maintenance Expenditures
OSS	One-Stop Shop
ppp	purchase power parity

PPP	Public-Private Partnership
SDA	Service Delivery Approach
SDM	Service Delivery Model
SWAp	Sector Wide Approach
Triple-S	Sustainable Services at Scale [IRC project]
UNICEF	United Nations Children's Fund
USA	United States of America
USAID	United States Agency for International Development
VLOM	Village Level Operation and Maintenance
WASH	Water, Sanitation and Hygiene

Benin

Commune	Local government unit
DG Eau	General Directorate of Water
PADEAR	Rural Water Supply Assistance and Development Programme
SONEB	*Société Nationale des Eaux du Bénin (National water utility of Benin)*

Burkina Faso

Commune	Local government unit
CREPA	*Centre Régionale pour l'Eau et l'Assainissement à faible coût* (network of resource centres as platforms for learning for state level officials)
ONEA	National Office for Water and Sanitation (responsible for urban areas)

Colombia

AQUACOL	*Asociación Colombiana de Organizaciones Comunitarias Prestadoras de Servicios de Agua y Saneamiento* (Colombian Association of Community-Based Water and Sanitation Services Providers)
CINARA	*Instituto de Investigación y Desarrollo en Agua Potable, Saneamiento Básico y Conservación de Recursos Hídricos* (research and development institute)
CRA	Water Regulatory Commission
MAVDT	Ministry of Environment, Housing and Territorial Development
PAAR	*Programa de Abastecimiento de Agua Rural* (Rural Water Supply Programme)
SSPD	Public Domestic Services Superintendent

Ethiopia

BoWRD	Bureaux of Water Resources Development
CDF	Community Development Fund
Kebele	Sub-district (municipality)
MoWE	Ministry of Water and Energy
MoWR	Ministry of Water Resources
WASHCO	WASH Committee
Woreda	District

Ghana

AVRL	Aqua Vitens Rand Limited
CONIWAS	Coalition of NGOs in Water and Sanitation
CWSA	Community Water and Sanitation Agency
DA	District Assembly
DiMES	District Monitoring and Evaluation
DWST	District Water and Sanitation Team
MLGRD	Ministry of Local Government and Rural Development
MMDAs	Metropolitan, Municipal and District Assemblies
MWRWH	Ministry of Water Resources, Works and Housing
RCC	Regional Coordinating Council
WATSAN	Water and Sanitation Committee
WSDB	Water and Sanitation Development Board

Honduras

AHJASA	Honduran Association of Water and Sanitation Boards
AJAM	Municipal Association of Water Boards
CONASA	*Consejo Nacional de Agua Potable y Saneamiento* (National Council for Water and Sanitation)
ERSAPS	*Ente Regulador de los Servicios de Agua Potable y Saneamiento* (regulatory body)
ESCASAL	*Escuelas y Casas Saludables* (Healthy Schools and Homes)
FHIS	*Fondo Hondureño de Inversión Social* (Honduran Social Investment Fund)
Mancomunidad	Association of municipalities
PEC	*Proyecto Ejecutado por la Comunidad* (community executed project)
RASHON	*Red de Agua y Saneamiento de Honduras* (Honduran water and sanitation network)

SANAA	*Servicio Autónomo Nacional de Acueductos y Alcantarillados* (National Autonomous Service for Water and Sewerage)
SIAR	*Sistema de Información de Acueductos Rurales* (Rural Water Supply Information System)
TOM	*Técnico en Operación y Mantenimiento* (Operation and Maintenance Technician)
USCL	Local Control and Supervision Unit

India

DDWS	Department of Drinking Water and Sanitation
DRD	Department of Rural Development
Gram Panchayat	Village level local government
GWSSB	Gujarat Water and Sewerage Board
IMIS	Integrated Management Information System (IMIS) of the Department of Drinking Water and Sanitation, Rajiv Ghandi National Drinking Water Mission
Pani Samitis	CBOs and/or sub-committees of local government responsible for water and sanitation (Village and Water Supply Committees)
PHED	Public Health Engineering Department
TWAD	Tamil Nadu Water Supply and Drainage Board
WASMO	Water and Sanitation Management Organisation, Gujarat State
WES-Net	Water and Environmental Sanitation Network

Mozambique

DAR	*Departamento de Água Rural* (Department of Rural Water)
DAS	District Water and Sanitation
DNA	*Direcção Nacional de Águas* (National Directorate for Water)
DPOPH	Provincial Department of Public Works and Housing
DWST	District Water and Sanitation Team
MPWH	Ministry of Public Works and Housing

South Africa

DWA	Department of Water Affairs
IDP	Integrated Development Plan
ISD	Institutional and Social Development (a training package for community mobilisation)

ANNEXES **165**

NSDP	National Spatial Development Perspective
MIG	municipal infrastructure
SSA	Support Services Agent
WSA	Water Services Authority
WSDP	Water Services Development Plan
WSP	Water Services Provider

Sri Lanka

CWSSP	Community Water Supply and Sanitation Project (formally ended in December 2010)
MWSD	Ministry of Water Supply and Drainage
NWSDB	National Water Supply and Drainage Board
Pradeshiya Sabhas	local government entities
RWSSC [1]	Rural Water Supply and Sanitation Centre
RWSSC [2]	Rural Water Supply Support Cell (in *Pradeshiya Sabhas* office)
RWSSU	Rural Water Supply Support Units (district-level)

Thailand

DLA	Department of Local Administration
DWR	Department of Water Resources
MOI	Ministry of Interior
MOPH	Ministry of Public Health
MWA	Metropolitan Waterworks Authority
PWA	Provincial Waterworks Authority
TAO	*Tambon* Administrative Organisation
RTG	Royal Thai Government

Uganda

CBMS	Community Based Management System
DLG	District Local Government
DWD	Directorate for Water Development (a department of the Ministry of Water and Environment)
DWO	District Water Office
DWSCC	District Water and Sanitation Coordination Committee
DWSDCG	District Water and Sanitation Development Conditional Grant

MWE	Ministry of Water and Environment
RGC	Rural Growth Centre
RUWASA	Rural Water and Sanitation
TSU	Technical Support Unit (deconcentrated office of the DWD)
UWASNET	Uganda Water and Sanitation NGO Network
WSSB	Water Supply and Sewerage Board (sub-county)
WUC	Water User Committee

USA

EPA	Environmental Protection Agency
NRWA	National Rural Water Association
RCAP	Rural Community Assistance Partnership
RUS	Rural Utilities Service
USDA	United States Department of Agriculture

ANNEX B
Glossary

Glossary	
Alignment	The process through which development partners align their aid to the recipient country's policy agenda and systems, such as their financial and monitoring systems.
Capacity support	The support activities towards water service authorities. They are typically provided by central ministries, or deconcentrated agencies of such ministries operating at regional or provincial level. It includes, among others, the provision of technical assistance, monitoring support and training of service authority staff.
Community-based management	The service provision option whereby communities control management of their water supplies. For practical purposes, day-to-day responsibility lies with a representative group of community people, often referred to as a water committee, elected to take up this task. Although this group may involve local caretakers or small entrepreneurs, the committee remains responsible for ensuring a sustainable service, and accountable to the community at large.
Coordination	In the context of aid effectiveness, the mechanisms (both formal and informal) through which sector actors articulate their activities and strategies amongst each other, and how they negotiate their role in or contribution to the sector.
Decentralisation	The transfer of authority and responsibility for governance and public service delivery from a higher to a lower level of government. There are different forms of decentralisation, as defined on p27.
Delegated management	All forms of contractual relationships between a water asset owner and an operator.
Harmonisation	The approach of donors coming together to develop common arrangements, procedures and information sharing mechanisms for their aid flows.
Intermediate level	The level where the functions of the service authority such as planning, coordination, regulation and oversight, and technical assistance, take place. We use the term the intermediate level (i.e. in between the national and community level) of local government, such as district, commune, governorate or municipality, or whatever the exact administrative name given in a particular country, as a generic term to describe this level.

Continued ▶

Continued

Life-cycle (of a water service)	The different stages through which a water services goes, from its initial capital investment phase, a service provision phase, the capital maintenance phase, and then subsequent upgrading, expansion and replacement.
Life-cycle costs	All the costs of water supply throughout its life-cycle. These include the categories as identified on p25.
Post-construction support	The ongoing support to water service providers, be they community-based or private. It may consist of aspects such as monitoring support, technical assistance, training and re-training, and advisory services.
Professionalisation (of community-based management)	The process of gradual involvement of professional staff in community-based service providers, and application of professional management principles to the service provider. In its simplest form, it may involve the hiring of a paid staff member, such as a plumber or administrator. More advanced forms may involve the hiring of an external operator to carry out some tasks. It also entails the application of principles of professional management, such as performance-based management, with the view of operating the service provision at professional standards.
Projectised	A term taken to mean a fragmented approach to investment or support which is largely disparate and un-coordinated and which results in a patchwork of solutions, rather than a more comprehensive or holistic approach
Project cycle	The cycle followed during the capital-intensive implementation phase of a water service.
Regulation	The provision of a set of rules, norms, monitoring and enforcement processes that ensure service providers meet nationally set guidelines and standards.
Scaling up	Scaling up is a familiar term in the water and sanitation sectors but means different things to different people. We refer to scaling up as the combination of vertical scaling up, or the institutionalisation of the functions and approaches that make sustainability possible, and horizontal scaling up, meaning the application of these principles in a broader geographical area.
Self-supply	The situation, in which individual households (or sometimes even a group of neighbours) invest in gradually improving their own service, and where the O&M is also done by the household themselves.
Service authority	Service authorities are the institutions that fulfil functions in relation to water supply, such as planning, coordination, regulation and oversight, and technical assistance, but not the actual service provision itself. Typically these authorities are located at the intermediate level and in most countries are carried out by local government (district, municipalities or communes).

Continued ▶

ANNEXES

Continued

Service provider	The institutions or individuals that deliver water to the users. They are responsible for the day-to-day provision of water, and include tasks such as operation, maintenance and administration of the water system. They may be community organisations, small private operators, public sector utilities or companies, or NGOs and faith-based organisations.
Sustainability	The concept is used liberally in the sector, and there are numerous interpretations of what this may mean. We follow the definition of Abrams (1998) describing sustainability as: 'whether or not something continues to work over time' (meaning, in this case, the indefinite provision of a water service [with certain agreed characteristics] over time).
Service Delivery Approach (SDA)	The conceptual approach taken at sector level to the provision of rural water supply services, which emphasises the entire life-cycle of a service, consisting of both the hard (engineering or construction elements) and software required to provide a certain service level.
Service Delivery Model (SDM)	The practical application of the principles behind the SDA to a given context, including agreed legal and institutional frameworks for delivering a service, the levels of service, and commonly understood and accepted roles for public, private or community actors.
Service levels	The normative set of attributes that describe the water service received. These typically include the quantity, quality, distance and continuity of the supply. These can be grouped into a service ladder.
Water service	The provision of access to a flow of water with certain characteristics, as defined in the service levels.

ANNEX C

Analytical tools for country studies	
Principle	**Explanation**
Enabling environment at national level	
1. Definition of Service Delivery Models (SDMs) and modalities in policy and laws	This element refers to the way in which water service delivery is formally defined in the national policy and legal framework, and the extent to which different sector stakeholders align to that. This includes, for example, a vision of the sector (targets and goals) and its broader position in development policy. A second aspect is the definition of the various levels of service (in terms of quantity, quality, distance, uses, provision to different sized settlements, functionality, etc.). Finally, this element refers to both the main paradigm(s) that exist regarding service delivery, and the modalities through which these can be provided, i.e. the definition of institutional framework for service delivery. Asset ownership is an important part of that; if there are doubts about where ownership lies, leveraging the financing for maintenance and asset replacement may be problematic.
2. Decentralisation policy	This element refers to the extent of and way in which decentralised service delivery is carried out in terms of the roles, responsibilities and resources, as well as the programmatic structures. For example, there may be one national water supply programme, guided from national level but carried out at decentralised level. Or, each local government may have its own programme. It also refers to the extent to which development partners contribute or not to this policy and programme. For countries where decentralisation is in process, it also refers to the way that process is structured and how decision-making, assets and staff are owned and/ or transferred to the decentralised level. Four facets of decentralisation are commonly seen: financial, political, functional and administrative.

Continued ▶

Continued

Principle	Explanation
Enabling environment at national level	
3. Oversight (regulation) and accountability	With decentralisation of responsibility for service delivery to intermediate levels, national government plays an increasingly important role in oversight, regulation and enforcement, so as to ensure accountability from service providers to users and to national governments, including elected branch of government. This element includes the frameworks, tools and mechanisms that have been put in place. As examples, sector monitoring and reporting at an aggregate level; innovative approaches to service provider accountability to national government; and the mutual accountability between governments and development partners.
4. Mechanisms for coordination, learning, support and technical assistance to intermediate level (sector learning)	In many countries decentralisation is not only about the formal policies and frameworks that guide it. Many local authorities need and will continue to need support in many forms, ranging from access to information, capacity to learn and reflect, technical assistance, etc. This element refers to the mechanisms that exist for learning and support, both at national level, and downwards to the intermediate level. It includes elements such as presence and use of sector information systems, resource centres, sector meetings, inclusion of water in university curricula, etc.
5. Financing	This element refers to four aspects: 1) the sources of financing (taxes, transfers, tariffs, donor funds, community contribution, private sector); 2) the way in which financial flows in the sector are earmarked, for example the percentage of grants to be dedicated to CapEx, OpEx, CapManEx, direct support costs, etc., but also what is needed at sector level for indirect support costs; 3) the ways in which these financial flows are coordinated and managed at national level (SWAps, five-year expenditure frameworks, off-budget, project-based), and downwards to the intermediate level (annual disbursement cycles, conditional grants, unconditional grants, project-based); and 4) an indication of the relative size of financial flows and routing.

Continued ▶

Continued

Principle	Explanation
Enabling environment at national level	
6. Organisational culture and behaviour with respect to harmonisation and coordination	This element refers to cultural and individual attitudes, experiences, beliefs and values of an organization at international, national and intermediate levels. The particular set of values and/or norms within groups and people in an organisation that direct the way in which they interact with each other and with stakeholders outside the organisation.
7. Institutional responsibilities for the different stages of the life-cycle of service provision	This element refers to the definition of roles and responsibilities for different functions (planning, construction, post-construction support, O&M, monitoring, training, etc.): who should fulfil the functions, and whether the functions are covered adequately by these actors.
8. Coordination mechanisms and platforms	Apart from a definition of the roles of each stakeholder in services provision, there is a need for coordination mechanisms between them. Under this element, the mechanisms (platforms, bodies, etc.) for such coordination are described and analysed in terms of their effectiveness. Coordination refers to all stages in the life-cycle, from coordination of efforts to address capital investment needs, to the identification of needs to provide post-construction support. Typical issues include coordination between NGOs active in a district; and mechanisms for coordination between those with governance functions and those with service provision functions. Coordination between different government bodies may also be included, particularly where some functions are decentralised and others are deconcentrated.
9. Monitoring and information systems for full service delivery	This element refers to mechanisms and systems in place for collecting a range of information on water systems (schemes) in districts, and access to these for use by different stakeholders in planning processes. It is also closely related to issues of access to information and accountability, both upwards to central government and donors, and downwards to communities.

Continued ▶

Continued

Principle	Explanation
Enabling environment at national level	
10. Strategic planning for full life-cycle for service delivery (capital projects, operations and post-construction support)	Under this element, the focus is on medium-term strategic planning approaches and mechanisms for the full life-cycle of delivery of services, according to the defined norms and standards, so entailing capital investments, ongoing provision and post-construction support for the entire area of jurisdiction at intermediate level. This also refers to how priority setting and targeting of investments is done for and with different groups within the area of jurisdiction. As examples, checking whether specific measures are in place to target the most vulnerable and poorest groups, whether there are pro-poor policies or criteria, and whether investments are biased to certain areas.
11. Financial planning for all life-cycle costs	This element refers to the financial component of strategic planning (see previous element). Such planning should consider all costs: CapEx, OpEx, CapManEx and direct support costs. It includes all income, and sources of income, including tariffs, transfers (from national government), taxes, donor grants, and both public and private investments. It also refers to the consistency between planning and availability of sources of funding (grants, direct investments, customer tariffs and contributions) to cover these costs, including both public and private financing mechanisms. Of particular importance is the clarity and consistency in terms of expected contributions of different customer groups and, inversely, the targeting of subsidies. Although this element is part of the previous one, it is listed as a separate element because of its importance, and the fact that it is often neglected.
12. Project implementation approaches	This refers to the approaches followed by actors at intermediate level, both in capital projects and ongoing support. Of particular importance is the standardisation of aspects such as creation of demand for improved services, health and hygiene promotion, and the use of supporting tools such as manuals and guidelines. Another aspect is how these approaches are articulated in short-term (annual) planning cycles, as well as in project cycles.

Continued ▶

Continued

Principle	Explanation
Enabling environment at national level	
13. Capacity (resources, supply chain, structures, systems and procedures, etc.) to fulfil functions during the entire life-cycle of service provision, and to carry out governance functions	Apart from clear responsibilities, there must be capacity at the intermediate level for both service provision and governance functions. Capacity refers to human resources (management, technical assistants, private operators, hardware shops, etc.) within the area, as well as material (computers, vehicles, etc.). The type of capacity required differs through the stages of the life-cycle and in terms of the type of system. For example, in the post-construction support phase spare part supply chains are relevant; during capital investment projects hardware and machines are needed, along with expertise in software.
14. Embedding delivery in framework for integrated water resources management (IWRM)	Sustainability of rural water supply services is affected more and more by increased competition over water resources. Rural water supply services therefore need to take into account water resources issues and the principles of IWRM. This implies that at levels above the community (sub-catchment, district, etc.) an assessment is made of available resources and how these affect service delivery. Both strategic planning at intermediate level and planning of capital works needs to be done within a framework for IWRM. In addition, efforts need to go into the promotion of representation of the rural water supply sector in platforms for water resources management. Under this element, an analysis should be made of how this is taken into account in service delivery. In many countries this implies looking at the interface between local government and water resources institutions.
15. Appropriate technology options	Technology options must be appropriate for the physical and socio-economic environment. Under this element, the focus is on the range of options available to communities to support full coverage, sustainability, and the ability to respond to changing demand for higher levels of service. A key issue is finding a balance between the development and use of innovative technologies, and standardisation to allow for economies of scale in, for example, the supply chain.

Continued ▶

Continued

Principle	Explanation
Enabling environment at national level	
16. Institutional arrangement for service provision	At community-level effective service providers need to be in place to manage the service. This can be CBOs, under the community-management approach, or other service provision management models (private operators, etc.). This element focuses on the type of providers that exist legally, as well as the type of contractual arrangements and regulations in place (service agreement, lease contract, etc.). Much of this should reflect national policy, but there is frequently local innovation and variation.
17. Mechanisms and approaches for customer participation in the full life-cycle of the service	The basis for sustainability is laid during capital works' projects. During such works, demand is created for services, and capacity is developed at community level to operate and manage the services. Ample evidence shows the importance of participatory planning tools and approaches in this. The same applies to other phases of the life-cycle. During O&M this can be in the form of mechanisms for customer relations and feed-back to service providers. Under this element the focus is on the mechanisms and approaches for customer participation, and the quality of these, during the full life-cycle.
18. Financial arrangements for service provision	This element looks at the financial arrangements for water services provision. A first aspect is clarity on expected customer contributions in different stages of the life-cycle, including initial contributions to capital works in the case of CBM, or other upfront investment arrangements. Another aspect is the arrangements in place for sound financial management, such as CBO bank accounts, access to commercial loans, billing software activities, or audits undertaken by independent auditors.

ANNEX D

Triple-S principles framework

Areas of principle	Levels of intervention			
	Water service provision	Intermediate	National	International
Policy, legislation and institutional factors	Water infrastructure, service levels and management arrangements are part of a recognised and defined Service Delivery Model (SDM), and do not operate in isolation.	Clear roles, responsibilities and authority exist at decentralised levels to ensure the delivery and oversight of water services under relevant management arrangements. They also exist for system construction, O&M, post-construction support, upgrading, system expansion and replacement.	Policies and institutional structures are adopted to enable the SDA. Service models, service levels and responsibilities for planning, regulation and providers are clearly defined. There is clear legal status for providers including asset ownership. Support is provided to all institutions responsible for service delivery at decentralised and service provider level.	Development partner funding policies support sector reform processes that enable the adoption of an SDA.
Financing	Service providers and consumers understand the benefits of full life-cycle costing. Clear strategies are in place to increase demand for a water service. There is a willingness to commit resources to operational and capital maintenance expenditure.	Financial planning accounts for full life-cycle costs, and service delivery is supported within available funding, through a combination of public sector financing, local revenues, tariffs and subsidies.	The concept of full life-cycle costs is embedded: financial mechanisms, budget processes and disbursement systems reflect this approach, including the costs of support to institutions at all levels. Total costs for service delivery are known and funded through a combination of national budgets, tariffs and (development partner) subsidies, as necessary.	Development-partner funding policies support full life-cycle costs, including non infrastructure elements, to enable an SDA.
Planning	Customers participate in planning processes and consultation mechanisms.	Planning at decentralised level is based on the SDA using economies of scale, with the aim of full coverage under appropriate management arrangements.	Planning at all levels is directed by clearly articulated policy choices and priorities, including concerns for IWRM and equitable access.	Development partner policies support decentralised planning processes.

Continued ▶

ANNEXES   177

Continued

	Areas of principle	Levels of intervention			
		Water service provision	Intermediate	National	International
Learning and adaptive capacity	Transparency and accountability	Customers have access to information and are informed about who is accountable for their water service; mechanisms are in place to enable them to voice their opinions on performance.	Instruments are enforced with adequate resources for oversight, monitoring and regulation of water service delivery, including tendering and contracting, as well as accountability to other stakeholders such as customers, providers and civil society.	Oversight, monitoring and regulatory instruments in place to ensure accountability of decentralised government for service delivery.	Development partner funding policies support adoption of transparency and accountability mechanisms at all levels.
	Awareness and skills	Service providers and customers are aware of their roles, rights and obligations, and have the skills and resources required to provide a sustainable service.	Skills, resources (including supply chains), information and long-term support to service providers are available at decentralised levels to ensure water governance functions.	Capacity building is a core policy with defined strategies, and is supported through investment.	Development partner funding policies support systemic investments to build capacity at all levels of the water sector.
	Culture of learning and information sharing	Service providers and customer stakeholders participate in reflection and debate around water service delivery at local and intermediate levels.	Support mechanisms are available with adequate resources to facilitate information gathering for learning and innovation to improve service delivery (including technologies and management arrangements).	A learning culture is encouraged at all levels. Resources and mechanisms are put in place to enable information sharing on sector performance and action research.	Development partner funding policies support the development of learning and innovation capacity in the water sector.
Harmonisation, alignment and coordination	Harmonisation and alignment	Water infrastructure design, technology and management arrangements adhere to national guidelines, norms, standards and approaches, regardless of the implementing entity.	Development partner-funded projects accept and work within planning, implementation and management arrangements, and within budget processes, monitoring arrangements and regulations for service provision set as part of the SDM.	Development partners are aligned with nationally-led policies, strategies, planning processes and priorities. Coordination mechanisms are in place for feeding development partner funding into the water sector.	Reciprocal accountability arrangements exist between national governments and development partners for rural water service policies and priorities. Development assistance is channelled through government-led mechanisms.
	Coordination		Coordination mechanisms and platforms are in place to apply the SDM and create economies of scale, both for construction of new systems and follow-up support.	Cooperation and integration between national ministries to ensure alignment of water and other sector policies.	Coordination between development partners is improved in support of the SDA.

ANNEX E
List of country reports and literature studies

All published reports are available for download at
www.waterservicesthatlast.org.

Adjinacou, C., 2011. Benin: L'alimentation en eau en zones rurales; evaluation des progrès vers la prestation de services durables. Unpublished report for IRC International Water and Sanitation Centre prepared by MGE Conseils.

Butterworth, J.B., 2010. *A brief review of service delivery concepts; a literature review*. The Hague: IRC International Water and Sanitation Centre.

Chaka, T. Yirgu, L. Abebe, Z. and Butterworth, J., 2011. *Ethiopia: Lessons for Rural Water Supply Assessing progress towards sustainable service delivery*. The Hague: IRC International Water and Sanitation Centre.

de la Harpe, J., 2011a. *Harmonisation and alignment literature review*. The Hague: IRC International Water and Sanitation Centre.

de la Harpe, J., 2011b. *South Africa: Lessons for Rural Water Supply; Assessing progress towards sustainable service delivery*. The Hague: IRC International Water and Sanitation Centre.

Gasteyer, S., 2011. *United States of America: Lessons for Rural Water Supply; Assessing progress towards sustainable service delivery*. The Hague: IRC International Water and Sanitation Centre and Michigan, USA: Michigan State University.

IRC and Aguaconsult, 2011. *Ghana: Lessons for Rural Water Supply; Assessing progress towards sustainable service delivery*. Accra, Ghana: IRC International Water and Sanitation Centre.

James, A.J., 2011a. *India: Lessons for Rural Water Supply; Assessing progress towards sustainable service delivery*. The Hague: IRC International Water and Sanitation Centre and Delhi: iMaCS.

James, A.J., 2011b. *Sri Lanka: Lessons for Rural Water Supply; Assessing progress towards sustainable service delivery*. The Hague: IRC International Water and Sanitation Centre and Delhi: iMaCS.

Juntopas, M. and Naruchaikusol, S., 2011. *Thailand: Lessons for Rural Water Supply; Assessing progress towards sustainable service delivery*. The Hague: IRC International Water and Sanitation Centre and Bangkok: Stockholm Environment Institute, Asia Centre.

López, M.A., 2011. *Honduras: Abastecimiento de agua en zonas rurales; Experiencias en la prestación de servicios sostenibles*. The Hague: IRC International Water and Sanitation Centre and Tegucigalpa, Honduras: RASHON.

Munguambe, C.G.J. and Langa de Jesus, V.A., 2011. *Moçambique: Abastecimento de água nas zonas rurais; Avaliação dos progressos para a prestação de serviços sustentáveis.* The Hague: IRC International Water and Sanitation Centre.

Nimanya, C., Nabunnya, H., Kyeyune, S. and Heijnen, H., 2011. *Uganda: Lessons for Rural Water Supply; Assessing progress towards sustainable service delivery.* The Hague: IRC International Water and Sanitation Centre and Kampala: NETWAS.

Rojas, J., Zamora, A., Tamayo, P. and García, M., 2011. *Colombia: Abastecimiento de agua en zonas rurales; Experiencias en la prestación de servicios sostenibles.* The Hague: IRC International Water and Sanitation Centre and Cali: Universidad del Valle/CINARA.

Zoungrana, D., 2011. *Burkina Faso: L'alimentation en eau en zones rurales; évaluation des progrès vers la prestation de services durables.* The Hague: IRC International Water and Sanitation Centre, and Ouagadougou: Institut International d'Ingénierie de l'Eau et de l'Environnement.

INDEX

access to water 19, 34, 36, 58, 60
accountability 27–8, 62, 65–69, 73, 85, 94–9
administration 19–20, 22, 25, 34, 80, 87
    see also local administration; public
Africa 16, 59, 63, 129, 131
    see also South Africa; West Africa
African Development Bank (AfDB) 94–5
agencies see CWSA; reforming agencies
agriculture 30, 32, 37, 54, 70, 132
aid
    dependency 1, 7, 15–7, 117–9, 121, 135, 137
    effectiveness 7, 27–8, 128–9, 131–5, 139
    see also WaterAid
alignment 8, 27, 53, 129, 135, 148
    see also harmonisation
alternative service providers 2, 83, 94, 143, 150
AQUACOL 35, 97, 105, 109–10, 117–8
asset
    holders 30, 74, 79, 83, 95–7, 119
    ownership 21, 49, 74, 76
    renewal 18, 26, 114–6, 122, 138, 146
asset management 2, 5–9, 19, 31, 108, 140–1
    planning 89, 116, 119, 134
assistance see development; RCAP; technical

Benin 15, 30–1, 119, 121, 128–9, 131
    decentralisation 66, 68, 123
    SDMs 74, 81–3, 87–9, 96, 105
    water supply 59–60, 64, 127
borehole 19, 22, 83, 93
bulk supply 26, 39, 42, 69, 128

Burkina Faso 5–6, 9, 12, 15, 32–3, 110
    decentralisation 67–72, 121–2, 125, 135
    expenditure 115–6, 119
    SDMs 74, 81–2, 86, 88–9, 97, 105
    water supply 59–60, 64, 127, 138, 148
    see also ONEA
business
    culture 35, 79–80, 96, 109
    development 83, 92, 116, 143
bye-laws 78, 86, 88, 95, 97, 113

capacity building 13, 21, 26, 55, 66, 96, 133
    finance 138, 151
    training 49, 52, 104, 106–7, 140
    see also sector capacity building
capacity support 123–5, 134, 139, 148
    post-construction 103–8
    service authorities 2, 8–9, 122, 144, 149–50
capital
    expenditure (CapEx) 20, 26, 113, 119–22, 146, 150
    investments 19, 36, 68–9, 119–22, 134, 145–8
capital maintenance 2–3, 7–8, 47, 55, 138–41
    expenditure (CapManEx) 26, 81, 113–22, 134, 146–9
centralisation 5, 44, 69–70, 89, 111–2, 142
Colombia 5, 9, 15, 34, 109–11
    decentralisation 67–9, 73–4, 125
    expenditure 114, 117–8
    SDMs 80–1, 89–90, 93, 96–100, 104–5
    water supply 58–9, 61, 139, 148
commune 21, 30–3, 67, 70, 81–2, 97

INDEX

community-based management (CBM) 1–3, 33–45, 73–87, 91, 96–107, 148–50
   professionalisation 4, 8–9, 21–2, 51–5, 116–8, 141–4
community-based organisation (CBO) 47, 63, 75–6, 123
   see also CWSA; CWSSP; Pani Samitis; RCAP
community development fund (CDF), Ethiopia 91, 123
community management 1–4, 30, 37, 78, 80, 142
   approaches 12–3, 17, 21, 24, 67–8, 88
community participation 1, 4, 62, 75, 85, 100
complex piped systems 4, 61, 68, 80, 82
compliance 54, 84, 97, 101, 123, 147
composite indicators 61–3
constraints 17, 51, 73, 94, 105
construction 12, 37, 43, 111–5, 127, 150
   costs 48–9, 113, 117–8, 138–40
   see also post-construction support; system construction
consumers 4, 23, 94–9, 113–7, 121, 147–9
   see also urban
continuity 11, 19, 58, 100, 119
contracts 47, 70, 81–7, 95–7, 135, 143
   see also lease contracts
coordination
   and harmonisation 6, 10, 57, 126, 133, 150
   mechanisms 41, 129, 131–2
corruption 6, 38, 145
cost
   categories 25–6, 114, 117, 119–22, 146
   see also construction; direct support costs; life-cycle; operation and maintenance; rehabilitation
coverage rates 5, 49, 111
   see also increased coverage; low coverage

country ownership 27, 129
customer care 46, 79, 92, 96, 109
CWSA (Community Water and Sanitation Agency) 37, 69–70, 77, 106, 117
CWSSP (Community Water Supply and Sanitation Project) 48–9, 107

Decade for Drinking Water and Sanitation 75, 142
decentralisation *see* Benin; Burkina Faso; Colombia; Ethiopia; fiscal decentralisation; Ghana; Honduras; India; Mozambique; South Africa; Sri Lanka; Thailand; Uganda; United States of America
decision making 17, 67–9, 88–90, 93, 129, 145
   power 25, 43, 67, 72, 80
deconcentration 21, 27, 44, 66
delegation 27, 29–30, 33, 51, 74–5, 78–82
delivery *see* service delivery
demand-driven 1, 44, 80, 105–6, 108–10, 123
demand-responsive approach (DRA) 44, 75, 107, 113, 142
Department of Water Affairs (DWA) 46, 70, 92, 97, 107, 123
dependency *see* aid; donor
design 18–9, 36, 43, 92–3, 128, 134
deterioration 19, 41, 63, 96, 117
development
   partner assistance 7–8, 148
   see also business; community development fund; MDGs; rural development;
direct support costs (DS) 7, 26, 35–7, 45, 89, 105–9
   see indirect support costs (IDS)
donor
   dependency 41, 121
   driven 1, 68, 115
   funding 96, 130–1, 139
   programmes 7, 45, 68–9, 73–5, 85, 117
drainage 48, 71, 106, 133
drinking water 42–4, 48, 54, 84, 101–3, 122

see also Decade for Drinking Water and Sanitation; improved water

economies of scale 5, 69–70, 81–9, 104, 111, 116
education 19, 37, 65, 90, 131
enabling environment 6–7, 17, 20, 22, 57, 111, 134
 functions 122, 140–1
enforcement 34, 84, 94–9
engineering 19, 42, 70–2, 89, 124
Ethiopia 5–6, 8, 15, 36, 100–2, 110
 decentralisation 67–9, 72, 121, 125–30, 132–3, 135
 expenditure 115, 119
 SDMs 74, 76, 85, 88, 91, 106
 water supply 59–60, 64, 138, 148
 see also community development fund; One WASH programme
expansion 11, 18, 25, 75, 81, 112
expenditure *see* Burkina Faso; capital; capital maintenance; Colombia; Ethiopia; Ghana; Honduras; Mozambique; replacement; South Africa; Thailand; Uganda

feasibility 18–9, 30, 62, 87, 92
federal 21, 36, 103, 108–9, 112–5, 132–3
 system 42, 54, 67
fees 39, 109, 117–8
finance 25, 32, 115, 120, 133, 151
 see also capacity building
financial
 flows 41, 119, 122
 frameworks 2, 112, 134, 146–7, 150
 management 27, 62, 79, 98
 mechanisms 8–9, 21, 55, 78, 140–1, 148–9
 planning 7, 21, 104, 108, 117, 134
 fiscal decentralisation 5, 27, 67, 70, 72–3
flows *see* financial; funding

fragmentation 6–8, 53, 69, 127–8, 135, 148
functionality 8–9, 60–4, 100–3, 126, 139, 148
 see also non-functionality; handpumps
funding
 flows 72, 119–20, 122, 131–3
 mechanisms 7, 37, 111, 114–7, 135
 see also donor
funds 53–5, 62, 68, 91, 107, 146
 see also community development fund; rotating funds *see* pooling

gender 76, 90, 101, 123
Ghana 9, 12, 15–6, 17, 38, 112
 decentralisation 66–70, 72, 123, 126–7, 130, 133
 expenditure 117, 119–20
 SDMs 74–7, 81–3, 86, 88–90, 101, 104–10
 water supply 59–60, 138, 148
Gini-coeficient 34, 46, 54
'golden indicators' 101–2
governance 16–7, 25, 48, 54, 73, 98
 functions 21, 65, 125
gram panchayats, India 42–3, 74, 78, 106
grants 72, 102, 114–7, 120–1, 123, 133
gravity fed 11, 53, 77, 87
guaranteeing 2–3, 34, 66–7, 88, 105
Gujarat 4, 15, 42–3, 67, 112, 141
 see WASMO

handpumps 53, 77, 82, 87, 133, 143
 functionality 11, 24, 58–64, 127
hardware 5, 26, 100, 110, 113–4, 128
harmonisation 12, 27, 37–9, 122, 144–5, 151
 and alignment 6–7, 16, 27, 111, 126–38
 see also coordination
health 19, 24, 37–8, 65, 98, 131–2
 see also public
higher service levels 4, 9, 59, 80–1, 139, 142–3

INDEX **183**

Honduras 6, 9, 15–6, 40, 96–103
    decentralisation 66, 69–70, 121–2, 124, 126–8, 130
    expenditure 112–4, 117
    SDMs 74–6, 80, 85, 90–3, 106, 108–10
    water supply 58–62, 139, 148
    see also TOMs
household connections 55, 58–9, 77, 80, 87
'hydraulic mission' 70–1, 139
hygiene 12, 37, 49, 90, 104
    see also WASH

implementation approaches 8–9, 49, 88–90, 138, 140, 144–51
improved water 38, 101
    service 1, 71, 83
    supply 46, 48, 54, 59
improvement 23, 69, 85, 98–102, 116, 151
income 15–17, 34, 46–50, 59, 80, 128
increased coverage 8–9, 11, 25, 47, 65, 147–8
India 5–6, 8–9, 101, 103–4, 133, 135
    decentralisation 71–2, 125–7
    SDMs 74–6, 79, 85, 91–4, 97, 106
    water supply 59–61, 63, 138, 148–9
    see also gram panchayats; Gujarat; Pani Samitis
indicators *see* composite indicators; 'golden indicators'; performance; proxy indicators; sustainability indicators; target indicators
indirect support costs (IDS) 2, 7, 26, 108, 117, 145–6
information
    management 99–101, 126
    sharing 39, 129
    systems 41, 63, 101–2, 123
INGOs (international non-governmental organisations) 28, 32, 68, 113, 130, 135
innovation 21–2, 88, 112, 135, 144
    see technology
institutional

    arrangements 24–6, 47, 57, 65–6
    frameworks 6, 9, 22, 54, 111, 149
investments *see* capital; sector
irrigation 42, 48, 71, 91, 109

Joint Monitoring Program (JMP) 11, 42, 58–9, 65

kebele 67, 129–30
knowledge management 2, 126, 144,

Latin America 5, 16, 59, 97, 109, 111
laws 25, 32, 76
    see also bye-laws
leadership 6–7, 33, 45–6, 112, 129–30, 135
learning 118, 135, 138, 144–7
    and sector capacity building 12–3, 57, 111, 122–9, 141, 144
    see knowledge management
lease contracts 31, 34, 95
legal
    frameworks 7, 21, 58, 111–2, 141–2, 149
    ownership 75–6, 143, 146
legislation 44, 47, 66, 82, 131, 149
    and policy 4, 86–8, 95, 111–3, 138, 140–3
life cycle 4–5, 69, 110, 120, 138–41, 144
    costs 2, 8, 113, 146–8, 150
    see also service delivery
loans 26, 49, 55, 115, 120
local administration 51–5, 107–11
    see Pradeshiya Sabhas
low coverage 5, 8, 84, 88, 110, 147

maintenance 22, 26, 63, 86, 108, 110
    backlog 7, 48, 122, 134
    see also capital maintenance; operation and maintenance (O&M)
Mali 16, 64, 68, 86
mancomunidades 5, 67–70, 106–9, 124
MDGs (Millennium Development Goals) 12, 33–6, 50, 59, 120, 122

Mozambique 5, 8, 15–6, 44–5, 97–8, 135
    decentralisation 63–4, 66, 69–70, 121, 125–8, 130–2
    expenditure 110–2, 119
    SDMs 72, 74, 88–90, 100, 106, 107
    water supply 59–61, 138–9, 147
municipalities 34–7, 40–6, 74, 79, 113, 117
    see kebele; mancomunidades

National Rural Water Association (NRWA), USA 55, 108–9
networks 11, 90, 120, 125–6
NGOs (non-governmental organisations) 37, 43, 53, 109, 120–1, 123
    and development 89–90, 125–6, 132, 135
    roles 76, 79, 84–6, 104–6, 129–30
    see also INGOs
non-functionality 1, 11, 43, 83
norms and standards 21, 46, 84, 92, 95, 150
not-for-profit 54–5, 75, 80

OECD (Organisation for Economic Co-operation and Development) 89, 121, 130–1
One WASH programme, Ethiopia 37, 53, 126, 129–30
ONEA (National Office for Water and Sanitation), Burkina Faso 32–3, 74
operation and maintenance (O&M) 19, 21, 40–5, 49, 52–3, 73, 100
    costs 37, 81–3, 106–8, 118, 120–1
    see also TOMs; VLOM
operators *see* private operators; rural
oversight 21, 51, 66, 68–70, 115, 141–3
    and enforcement 93–9
    of services 9, 43–4, 92, 147–8, 150
ownership 33, 54, 66, 70, 79, 89–91
    see also asset; country ownership; legal

Pani Samitis, India 42, 78, 106
participation 73–5, 90, 123, 126, 142
    see also community participation
partnerships 4, 12, 81, 94, 129, 131–2
    see also development; PPPs; RCAP
performance
    indicators 64, 126, 96, 139, 143
    management 65, 80, 100–1
    see also poor
peri-urban 32–3, 83
piped systems 11, 44, 50–2, 84–8, 127–8
    see also complex piped systems; small piped systems
point source 4, 30–1, 76–7, 87, 143
political
    economy 6, 145, 151
    engagement 6, 10, 27, 145, 151
pooling 33, 116, 124, 129
poor
    performance 11, 63, 99, 102, 111, 132
    sustainability 11–2, 24, 62, 90
population growth 11, 61, 81, 87, 116, 143
post-construction support 41, 44–5, 49, 69–70, 134, 138
    activities 24–6, 96–9
    arrangements 103–11, 146
    mechanisms 76, 80, 88–91, 96, 117–8
    to service providers 1–8, 35, 55, 104–9, 140–4, 148–50
    see also capacity support
potable water 36, 54, 62
poverty line 32, 34, 40, 44, 46
PPPs (public-private partnerships) 52–5, 74–5, 85–7
Pradeshiya Sabhas, Sri Lanka 48–9, 107
private operators 30, 78, 91–4, 98, 105, 142–4
    see also PPPs; small-scale private operators
professionalisation 2, 13, 31, 77, 80–8, 150

INDEX

promotion 68, 80, 85, 123, 126, 150
 see alternative service providers
protection 43, 54, 85, 93–4
provision *see* service provision
proxy indicators 60, 65
public
 administration 5, 73, 135
 sector 3, 22, 71–3, 80–1, 87, 125–7
 works 44, 50
 health 40–2, 50, 54, 70, 123
 see also PPPs
pumps 19, 23, 26, 83, 133
 see also handpumps

quality
 see water quality
quantity 18, 22, 58, 101

rainwater harvesting 43, 49, 51, 75, 84–7, 94
RCAP (Rural Community Assistance Partnership), USA 55, 108–10
reforming agencies 70–3
reforms 13, 38, 65–73, 112, 130, 145–9
regulation 21, 24, 30–1, 54–7, 92–100, 123, 139–41
 of rural services 2–3, 69, 84–8, 110, 147–8, 150
rehabilitation 34–5, 47, 89, 118, 139–40, 150
 costs 26, 55, 114, 120, 146
reliability 22, 58, 63, 100, 103
renewal *see* asset
replacement 18, 20, 73, 89, 116, 119, 138
 expenditure 7, 25–6, 55, 122, 134, 140–1
resistance 3, 69–72, 112
reticulated systems 77, 87, 143
revenue 26–7, 50, 81, 117
 collection 4, 46, 87, 92, 143
RGCs (Rural Growth Centres) 52–3, 68, 75
river basin 42, 91, 93
rotating funds 8, 148
'rules of the game' 6–7, 97, 134, 138, 141, 149

rural development 38, 42, 133
 see also RCAP; RGCs
rural operators 2, 35, 55, 94–9, 109, 147
rural water services 9, 25, 57–60, 67–73, 78
 at scale 12–6, 137, 147
 see also regulation
Rwanda 81, 86, 130

sanitation *see* AQUACOL; water and sanitation
scalability 6, 76, 138
scaling up 16, 24–5, 37, 140–1
SDA (service delivery approach) 17–22, 71–2, 88, 110–1, 130–41, 147–9
 sustainability 41, 65
SDMs (service delivery models) *see* Benin; Burkina Faso; Colombia; Ethiopia; Ghana; Honduras; India; Mozambique; South Africa; Sri Lanka; Thailand; Uganda; United States of America
sector
 change 10–1, 149–50
 investments 119–20, 128
 see also public; SWAp; WASH
sector capacity building 111, 122, 125
 see also learning
self-supply 3–4, 9, 22, 103, 143, 148
 approach 37–45, 51–60, 73–5, 83–7
Senegal 16, 68, 81, 109
service
 ladder 22–3, 79
 levels 22–4, 58, 87, 99, 134, 150
 -orientated
 see also improved water; post-construction support; rural water services; WSA; WSDBs
service authority functions 2–5, 66–9, 95–7, 110–1, 139, 144
 intermediate level 21, 24, 57, 88
service delivery
 life-cycle 2–3, 19–20, 25, 73, 92, 134
 see also SDA; SDMs

service provision 12, 40, 55, 66–9, 72–3, 99–100
  models 34–5, 47–8, 86–8
  options 16–7, 22, 24, 57–8, 75, 79–80, 107
  see also alternative service providers; WSPs
sewerage 40, 53, 116
skills 71, 81, 105–6, 123, 131, 140–1
slippage 9, 23, 42–3, 55, 60–3, 138, 148
small piped systems 30, 38, 58–60, 77, 88, 127
small-scale private operators 1–4, 68, 72, 81–6, 118, 149–50
small towns 17, 30, 36–9, 55, 69, 81–2
software 26, 62, 79, 90, 113–4
  activities 5, 19, 114, 120
South Africa 5–7, 9, 16, 46–7, 97–9, 134–5
  decentralisation 67–8, 70, 122–4, 129
  expenditure 111–2, 114, 117–9
  SDMs 72–4, 80–4, 89–94, 104, 107–8
  water supply 59, 139, 149
springs 53, 60, 87, 93
Sri Lanka 8, 15, 48–9, 59, 148
  decentralisation 70–1, 123, 128
  SDMs 74, 79–80, 104, 107
  see also Pradeshiya Sabhas
stakeholders 16, 26–8, 30–3, 47, 66
standardisation 127–9, 144, 150
standards see norms and standards
state level 22, 42–3, 54–5, 72, 106–8
storage 18, 51, 87, 101
subsidies 85, 114–6, 134
supply
  chains 37, 76, 107, 127–8
  -driven 1, 108, 110, 124
  see also bulk supply; self-supply; water supply
support see capacity support; post-construction support; TSUs
surface water 93, 128
sustainability indicators 2, 63–5, 143, 150
  see also targets for sustainability
Sustainable Services at Scale see Triple-S
SWAp (Sector Wide Approach) 28, 39, 45–7, 92, 112, 119
  mechanisms 53, 64, 130–5, 139
system construction 19–20, 26, 41, 70, 83, 90

Tamil Nadu 15, 71, 106, 133
target indicators 2, 150
targets for sustainability 43, 57, 64, 122, 144
tariff 21, 37, 77–81, 96–8, 114, 118
taxes 113–7, 146
technical
  assistance 21, 34–5, 54, 85, 118, 144
  support 36, 50–2, 79, 107–10, 123
  see also TSUs
technicians 26, 55, 62, 96, 105–7
  see also TOMs
technology 2, 39, 49, 74, 82, 87–8
  and innovation 125, 127–8, 140–1
  options 53, 58, 127–8, 143
tender 31, 70, 92, 145
Thailand 4–6, 8, 15, 50–1
  decentralisation 68, 127
  expenditure 111–2
  SDMs 75, 78, 84–5, 98, 104, 107
  water supply 59, 139, 148
TOMs (operation and maintenance technicians), Honduras 96, 98, 102, 108–10, 117
town see small towns
training 3–5, 79, 90–6, 123–5, 138, 144
  see also capacity building
transparency see accountability
Triple-S 1, 12, 28, 141
TSUs (Technical Support Units), Uganda 52, 81, 89, 108, 117, 122–3

Uganda 5–7, 12, 15, 17, 52, 97
  decentralisation 64, 67–8, 119–23, 125–7, 129–30

expenditure 112, 114, 116–7
   SDMs 72–5, 81–5, 89–90, 93, 102–4, 108
   water supply 59–60, 132, 134–5, 139, 148–9
   see also TSUs
United States of America 5, 8, 15, 54, 115–7
   decentralisation 65, 67–8, 121–2
   SDMs 76, 81, 85, 91, 98, 108–11
   water supply 59, 132, 134, 138–9, 148
   see also National Rural Water Association; RCAP
urban
   consumers 44, 61, 80–1
   utilities 7, 35, 73–9, 105–9, 118, 134
   water supply 38, 91, 96–8
   see also peri-urban
USAID 41, 117, 130
user *see* water user
utilities 3, 22, 54–5, 89, 116, 146

village level 44, 48, 111, 133
VLOM (village level operation and maintenance) 44–5, 68, 75, 79
voluntary 41, 51, 110, 118, 142–3, 150
   arrangements 4, 33, 76, 85–7
VWCs (village water committees) 50–1, 84, 107

WASH (Water, Sanitation and Hygiene) 12, 46, 65, 101, 141
   committees 37, 106
   sector 27, 111, 120–3, 128–31
   see also One WASH Programme
WASMO (Water and Sanitation Management Organisation) 4–5, 43, 71, 78, 106–7, 112
water
   boards 80, 37, 42, 80, 98, 101
   committee 4, 21, 61–2, 80–5, 96–7, 149–50

   development 13, 52, 123, 133, 139
   facilities 52, 60–1
   infrastructure 11, 113, 116, 145
   ministries 36, 42, 48, 52, 105–6, 120
   point 23, 44, 58, 61, 63, 101
   policy 36, 46, 92
   scarcity 91, 93
   see also access to water; improved water; potable water; rural water services; surface water; water and sanitation; water quality; water resources; water supply;
WaterAid 11, 72
water and sanitation 33–40, 54, 68, 90, 114–5, 125–6
   see also CWSA; CWSSP; Decade for Drinking Water and Sanitation;
water quality 36, 61–2, 80, 87, 104, 107
   regulations 54–5, 58, 98–101
water resources 17, 28, 32–8, 42–3, 107, 137
   management 50–2, 69–71, 88, 91–3, 101, 103
water supply *see* Benin; Burkina Faso; Colombia; CWSSP; Ethiopia; Ghana; Honduras; India; Mozambique; South Africa; Thailand; Uganda; urban
water user 31, 52, 74, 82, 101, 104–5
wells 37, 51, 55, 60, 83–5, 87
West Africa 30, 32, 68, 81, 86
woredas (districts) 36–7, 67, 106, 116, 129, 132
World Bank 24, 32, 48, 112, 130
WSA (water services authority) 47, 68, 92–4, 97, 107, 124
WSDBs (water and sanitation development boards) 39, 70, 77–8
WSPs (water services providers) 46–7, 68, 83–6, 92, 97, 107

IRC International Water and Sanitation Centre, The Netherlands

IRC is a modern NGO that is focused on working with a worldwide network of partner organisations as an innovative knowledge centre. IRC's roots are in advocacy, lesson learning, knowledge management and capacity building. IRC has built up a strong reputation for cutting-edge innovation and action research aimed at providing equitable and sustainable WASH services at scale.

IRC does not directly provide WASH services. Rather we work with those who do, helping them to tackle the systemic challenges that lead to wasted investment; thereby making their investments more effective in delivering services to all citizens. This means working actively with practitioners in a number of focus countries in a participatory action research mode to identify challenges, fill gaps, and develop robust models for sustainable and equitable service delivery. We take the lessons learned from these activities and use them to advocate at national and international levels for improved policy and practice. And, we document the results of these experiences and share them with a wide international audience.

IRC is leading two major initiatives to improve sustainability in WASH services funded by the Bill & Melinda Gates Foundation: Triple-S (Sustainable Services at Scale) and WASHCost.

For more information, please go to www.irc.nl

Aguaconsult, UK

Aguaconsult is a UK-based company providing technical assistance and consulting inputs for clients around the world and has developed capacity and a proven-track record in relation to two principal fields: water supply, sanitation and hygiene and disaster risk reduction.

We work in partnership with client organisations from the public, private and non-profit sectors, providing collaborative solutions for a range of needs – from the design of sectoral or thematic programmes to evaluations and research. We draw on a network of professional consultants to follow-through with specialised services, including expert technical assistance, operational support and technical back-stopping. The company has expertise in all levels of intervention – from community-based initiatives to research and policy development. It has geographic experience across Latin America, Africa, the Central Asian states, and South and Southeast Asia.

For more information, please go to www.aguaconsult.co.uk